Venezuela's Petro-Diplomacy

UNIVERSITY PRESS OF FLORIDA

Florida A&M University, Tallahassee
Florida Atlantic University, Boca Raton
Florida Gulf Coast University, Ft. Myers
Florida International University, Miami
Florida State University, Tallahassee
New College of Florida, Sarasota
University of Central Florida, Orlando
University of Florida, Gainesville
University of North Florida, Jacksonville
University of South Florida, Tampa
University of West Florida, Pensacola

# Venezuela's Petro-Diplomacy

Hugo Chávez's Foreign Policy

Edited by Ralph S. Clem and Anthony P. Maingot

FOREWORD BY CRISTINA EGUIZÁBAL

UNIVERSITY PRESS OF FLORIDA

Gainesville · Tallahassee · Tampa · Boca Raton

Pensacola · Orlando · Miami · Jacksonville · Ft. Myers · Sarasota

Copyright 2011 by Ralph S. Clem and Anthony P. Maingot

First cloth printing, 2010
First paperback printing, 2015

Library of Congress Cataloging-in-Publication Data
Venezuela's petro-diplomacy : Hugo Chávez's foreign policy/ edited by Ralph
S. Clem and Anthony P. Maingot ; foreword by Cristina Eguizábal.
p. cm.
Includes index.
ISBN 978 0-8130-3530-7 (cloth: alk. paper)
ISBN 978-0-8130-6142-9 (pbk.)
1. Venezuela—Foreign relations. 2. Chávez Frías, Hugo. 3. Oil industries—
Venezuela. 4. Petroleum industry and trade—Venezuela. I. Clem, Ralph S. II.
Maingot, Anthony P.
JZ1561.V46 2011
327.87—dc22
2010031993

The University Press of Florida is the scholarly publishing agency for the
State University System of Florida, comprising Florida A&M University,
Florida Atlantic University, Florida Gulf Coast University, Florida
International University, Florida State University, New College of Florida,
University of Central Florida, University of Florida, University of North
Florida, University of South Florida, and University of West Florida.

University Press of Florida
15 Northwest 15th Street
Gainesville, FL 32611-2079
http://www.upf.com

# Contents

# Figures and Tables

# Foreword

Unlike the governments of some other oil-rich states, Venezuela's democratic governments have historically been extremely adept at using oil and the wealth derived from oil exports as a foreign policy tool. None has been bolder than the government of Hugo Chávez, Venezuela's current president. Not only has Chávez established his country as a key player in inter-American relations, he has also crafted for himself a leading role worldwide.

With oil to back him, President Chávez frames his foreign policy according to the historical Bolivarian ideals of continental sovereignty and solidarity, just as prior Venezuelan leaders have done: Rómulo Betancourt sought support for the country's nascent democracy; Carlos Andrés Pérez pursued a new world economic order; and Rafael Caldera favored the cause of international social justice. This time, the Bolivarian dream is the inspiration for twenty-first-century socialism and the excuse for harsh anti-U.S. rhetoric.

To better understand the dynamics of Venezuela's international relations, the Inter-American Program (formerly known as the Summit of the Americas Center) at Florida International University's Latin American and Caribbean Center (LACC), with the support of the university's Western Hemisphere Security Analysis Center (especially Brain Fonseca and Jerry Miller), convened a group of Venezuela specialists, academics, and other international observers for a day of analysis and exchange of views on the inner logic and outer impacts of Venezuela's foreign relations policy. The resulting conference, "Ten Years of Venezuelan Foreign Policy: Impacts on the Hemisphere and the World," was held in Miami on May 29, 2008. This volume is an updated compilation of the presentations discussed at that conference, complemented by contributions from key analysts of Venezuela's current affairs.

This project would not have been possible without the assistance of Carl Cira and Elisa Gallo, who have since left FIU. They worked hard to identify the participants and set the agenda, and they proposed to me the idea of publishing the conference proceedings. I am indebted to two of FIU's top senior scholars, Ralph S. Clem and Anthony P. Maingot, who accepted the challenge of putting together this very interesting and, we hope, timely collection of articles. Pedro Botta, then of LACC and now on the staff of FIU's School of Public and International Affairs (SIPA), was instrumental in various aspects of manuscript preparation and in assisting the editors in coordinating the project. Finally, I would like to acknowledge Alisa Newman for her skillful copyediting of the original manuscript submissions and for the translation of the chapters by Román Ortiz.

*Cristina Eguizábal*
*Director, Latin American and Caribbean Center*
*School of International and Public Affairs*
*Florida International University*

# Introduction

## Continuity and Change in Venezuela's Petro-Diplomacy

RALPH S. CLEM AND ANTHONY P. MAINGOT

Over the last decade, Venezuela has exerted an influence on the Western Hemisphere and indeed, global international relations, well beyond what one might expect from a country of 26.5 million people. In considering the reasons for this influence, one must of course take into account the country's vast petroleum deposits. The wealth these generate, used traditionally to finance national priorities, more recently has been heavily deployed to bolster Venezuela's ambitious "Bolivarian" foreign policy.

As with any country whose national income depends so heavily on export earnings from a single natural resource, Venezuela's economic fate rests on its capacity to exploit that resource and, perhaps even more so, on the price set by the international market. The Venezuelan author Fernando Coronil calls this condition the "neo-colonial disease." Without oil, there would be no Chávez, and certainly no "socialism of the twenty-first century."[1] Unpredictability is the signal character of the Venezuelan economy, and, consequently, no domestic or foreign policy assessments can be completely time-based. Nothing exemplifies this volatility better than the price of a barrel of oil, which was $148 when we began planning this collection of essays in mid-2007, but which fell to less than $50, then climbed back to $70 by mid-2009. Instead of using a time-based approach, therefore, the essays in this volume attempt to trace the deep-rooted and enduring orientations of Venezuelan political culture and their influence on foreign policy, with special attention to the historical role petroleum has played in these considerations.

## The "Bolivarian" Ideal

Venezuela's sense of itself as having a "continental" or hemispheric role is not a new phenomenon. From the very inception of the Wars of Liberation in 1810, Simón Bolívar called for a confederation of states in the Southern Hemisphere. In 1815, while in exile in Jamaica, he spelled out the geopolitical reasons that made Venezuela a natural pivot of hemispheric unity: "The Americans have come to know each other . . . because of their physical geography, the vicissitudes of the war and the calculations required by the war."[2] Attempting to elicit British support, Bolívar developed his own theory of international relations, which, as John Lynch writes, encompassed "a total vision of America, beyond Venezuela and New Granada."[3]

Notwithstanding his exceptional personal qualities and merits, however, Bolívar could not stop the centrifugal forces that destroyed his efforts at a "grand confederation." One month before his death in 1830, he wrote a letter to the president of Ecuador lamenting that twenty years of effort to achieve his hemispheric ideal had been in vain. His final words were dramatic and fatidical: "Our America is ungovernable. Those who serve a revolution are plowing the sea." The only thing to do in America, he concluded, is to emigrate. He was bitter about the role of the *caudillos* in engendering "primitive chaos" and, consequently, in preventing any and all ideals of hemispheric unity. His thoughts on what we now call *caudillismo* and populism were prophetic:

> Unfortunately, among us, the masses can do nothing, a few strong wills do it all and the multitudes follow their audacity without examining the justice or the crimes of the *caudillos*; then they abandon him when one even more perfidious surprises him. This is the nature of public opinion and national character of our America.[4]

Bolívar was not mistaken. Like so many other nations in the hemisphere, Venezuela in the nineteenth century was a society at war with itself. As *caudillos* of the Conservative and Liberal oligarchies battled each other, European navies blockaded the nation more than once, demanding compensation for damages or the repayment of debts owed to their nationals.

Rampant *caudillismo* was subdued, though not completely eradicated, with the rise to power of Juan Vicente Gómez (1908–35). Gómez presided over two fundamental changes in Venezuelan society: the formation of a centralized, professional military, and the establishment of an oil industry controlled by foreign interests. As in Mexico, where another dictator, Porfirio Díaz, had invited foreign interests to exploit his nation's oil, Venezuela's relations with other nations now sprang fundamentally from the role of U.S. and European companies on her territory.[5] With the United States already importing oil, the *New York Times* celebrated the Gómez regime as "the arrival of a Venezuelan Díaz."[6]

From that point on, Venezuela's international relations were nearly completely centered on those nations with interests in the oil industry, and on their colonies. Trinidad, Curaçao, and Aruba were of special concern for two reasons: the oil refineries were located there, and opponents of the regime often sought refuge on these islands, from which they launched invasions. These interests aside, however, and despite the growing importance of the oil economy, Venezuelan political culture, in thought and action, focused on domestic governance—that is, on controlling both the new and the old centrifugal forces that kept Gómez's secret police perpetually alert and deadly. The philosophical underpinnings of that iron-like rule were provided by what Laureano Vallenilla Lanz characterized as "Cesarismo democrático." According to Vallenilla, the "true character" of Venezuelan democracy was the predominance of a determined individual whose power was based on "the wishes of the great popular majority, tacitly or explicitly expressed"; in other words, a populist *caudillo* whose preoccupations were purely domestic.[7] Hemispheric or other international matters were not of great concern during this period.

Venezuela's interest in hemispheric foreign relations reemerged with the birth of democratic politics and the election of Rómulo Gallegos of the Acción Democrática (AD) party in 1948. The first item on the new government's agenda was the revision of the petroleum law to institute "50/50" profit-sharing, an arrangement that was soon copied by oil producers worldwide. Venezuela, as distinct from Mexico, visualized an international role for itself in petroleum politics. Alas, just as Bolívar had anticipated, that other strain of Venezuelan politics, *caudillismo*, sprang back into action. Gallegos was overthrown, and a military regime governed until 1959, when new elections returned Acción Democrática to power under Rómulo Betancourt. Betancourt can be credited with reinitiating Venezuela's hemispheric-wide—that is to say, "Bolivarian"—foreign policy. As his biographer points out, Betancourt was not only a Venezuelan leader, he was also a hemispheric leader—an "Americanist in the Bolivarian concept of Latin American unity."[8]

The deep Bolivarian strand in Venezuela's political culture had not vanished, and neither had the idea of Venezuela as the pivot of a hemispheric "revolution." Betancourt expressed these ideals explicitly when he wrote of his hope for a wide-reaching Venezuelan foreign policy in a hemisphere "in revolution":

> Venezuela's socio-political process cannot be separated from what we find in all the other countries of Latin America. Our America has "entered into revolution," to use [José Martí's] words. From one extreme of the continent to the other one notes a swift tide of popular insurgency. . . . It will fall to Venezuela to play an important role in Latin America's integration process.[9]

Those who have studied Venezuela's foreign policy argue that Betancourt was simply picking up the Bolivarian doctrine of a hemispheric foreign policy: "Venezuela's diplomacy was born under the sign of the continent."[10] The same can be argued of Hugo Chávez. According to Harold Trinkunas, Venezuela's recent actions on the global stage stem from this longer tradition of oil diplomacy; the actors may have changed, but the core rationale remains the same.[11] Trinkunas develops this theme further in his chapter in this book.

Interestingly enough, during the 1960s the United States tolerated Venezuela's increasingly aggressive steps to take a larger share of its oil profits. Certainly Venezuela's hostility toward Fidel Castro's Cuba had something to do with this tolerance, on the principle that "the enemy of my enemy is my friend." Perhaps more important, however, was the growing U.S. dependence on Venezuelan oil. Arthur Schlesinger Jr. emphasized this point when he explained why President Kennedy was the first U.S. leader to visit Venezuela. "Some of us in Washington," he noted, "saw Venezuela as a model for Latin America's progressive democracy," adding (in parentheses), "remembering always that its oil revenues gave it a margin of wealth the other republics lacked."[12] This was petro-diplomacy at its best. It differed from the present phase of petro-diplomacy in that neither Betancourt nor the United States phrased their interest in the oil industry by using hostile geopolitical rhetoric. Betancourt had undergone an ideological change of heart; after flirting with an incipient communist movement in his youth, he had opted for, as he put it, the "West" and its democratic system.

Following two periods of AD government, the Bolivarian tilt of Venezuela's foreign policy experienced one of its periodic down cycles with the victory of the Christian Democratic Party (COPEI). It had reemerged with new vigor under the AD government of Carlos Andrés Pérez in 1974. Again, petro-politics began with the nationalization of the oil industry. Just as the 50/50 royalty policy was emulated by other oil producers, Venezuela's nationalized oil company (Petróleos de Venezuela, S.A., or PDVSA) became a model for state-owned companies elsewhere. By this time, Venezuela was the world's third largest exporter of oil and the most important source of crude oil imports into the United States.

Pérez used the country's growing oil wealth to resuscitate the Bolivarian dream of hemispheric unity. The list of Venezuela's financial contributions to initiatives in Latin America and the Caribbean under his regime provides a comparative perspective for current efforts to pursue a Bolivarian foreign policy:

$500 million to the World Bank
$500 million to the Inter-American Development Bank
$500 million to the International Monetary Fund
$60 million to the Andean Development Corporation
$25 million to the Caribbean Development Bank
$80 million to the Central American Coffee Marketing Corporation[13]

Pérez's foreign policy initiatives were not limited to financial assistance; Venezuela also developed several geopolitical initiatives that irritated the U.S. government. One of these was an effort to legitimize ideological pluralism within the traditionally anti-communist hemispheric defense system known as the Rio Treaty, as part of a larger effort to reintegrate Cuba into the Organization of American States. Similarly, Venezuela's support and funding of the Sistema Económico Latinoamericano (SELA) specifically included Cuba but excluded U.S. membership. Pérez also backed Panama's position in that country's negotiations with the United States over the Panama Canal, a hot-button issue in U.S. domestic politics at the time. And much as it may have annoyed the U.S. administration, Pérez's open hostility toward the Pinochet regime in Chile was consistent with the "Betancourt doctrine" of opposing dictatorships (with the exception of Cuba).

By 1976, Venezuela's oil-driven foreign policy initiatives had run out of steam. They came up against the same two powerful hemispheric forces that had stymied Bolívar a century and a half before: fragmentation and nationalist rivalry.[14] Equally significant, domestic politics, always turbulent, were now rocked by arguments over the distribution of oil wealth. "Venezuela," wrote one scholar of the oil industry in 1977, "is a vibrant, rough-and-tumble democracy, and it is unrealistic to expect contenders for power to grant a special status to the oil industry leaving it out of their struggles."[15] In this volume, John Magdaleno explains how this domestic reality operates in contemporary Venezuela. Then, as now, oil wealth has led to what John Martz called "bursts of impetuosity" and "impractical and ill-considered" foreign policy decisions, which, in turn, have soured the mood of domestic politics.[16] It was so under Pérez and it is so under Chávez, with one major qualitative difference that is worth repeating: the harsh anti-U.S. rhetoric and foreign policy initiatives of the Chávez regime make the slight tilt to the left of the Pérez period pale by comparison. In November 2008, Simón Romero of the *New York Times,* describing Caracas as a "hub for leftists," interviewed many of the two hundred Marxists enjoying an all-expense-paid stay in the city's finest hotels. "Is Chávez a mere populist or a genuine revolutionary?" an Egyptian socialist asked. "I dismiss the first idea."[17]

And yet, for all the talk of continuity in petro-state politics, there is a major difference in the domestic use of oil profits by the Chávez regime and their use by previous administrations. "Sowing the oil" had been the goal of the Venezuelan state since at least 1940. This has meant investing in the creation of steel, aluminum, and petrochemical industries, all geared toward creating export markets. Under Chávez's nationalization programs, this production is geared toward the domestic market, making the "Magical State," as Coronil calls it, completely dependent on oil as a source of foreign exchange. "If the price of oil were to falter," a former Minister of Planning cautions, "Venezuela would have no alternative industries that could expand to take its role in generating foreign exchange."[18] This certainly is not the view of the Chávez regime. As late as November, 2008, it showed little concern (publicly) over the global recession. Venezuela, said then Minister of Finance Ali Rodríguez, "is one of the most stable economies in the world." All this led *The Economist* to declare that Venezuela was "in denial" about the state of its economy.[19] In fact, by mid-2009 Venezuela was experiencing the highest inflation rate in the hemisphere, a slide in private-sector confidence and intentions to invest in the future, and a growing shortages of staples.[20] These shortages were certainly aggravated by the tensions between Venezuela and Colombia, with Chávez declaring that "Venezuela will reduce its trade with

Colombia to zero."[21] Certainly, with exports totaling $6 billion in 2008, Venezuela is Colombia's second-largest export market after the United States. Because agricultural products make up the bulk of those exports, this situation has led a highly regarded think tank to warn that a total cutoff of food item exports in Venezuela would provoke an immediate and angry response from the public. In addition, it would come "at a time when domestic political opposition in Venezuela is running high, with the economy in dire straits and disparate parts of the public in an uproar about Chávez's plan to launch a new and nationwide pro-regime education program."[22] A similar set of contradictions characterize Chávez's economic relations with the United States, its largest market. As an example, fully $33.4 billion of PDVSA's total sales receipts of $122.5 billion derive from the PDVSA-owned CITGO refining and retail gasoline stations in the United States.[23] It is a fact, however, that U.S. dependence on Venezuelan oil has been decreasing. Before Chávez came to power, the United States imported 17 percent of its oil from Venezuela; by 2008, that had been reduced to 9.6 percent.[24]

As impractical and ill considered as the Chávez domestic policy might be, it is Chávez's foreign policy initiatives which, to revisit the language of John Martz, illustrate the most telling "bursts of impetuosity." María Teresa Romero and Ralph Clem provide vivid examples of this foreign policy style in their chapters, and Roman Ortiz addresses its impact on Colombian security, both military and economic.

## The Global Context

In July 2006, Chávez gave the opening address at the African Union Summit in Banjul, The Gambia. As counterintuitive as this might seem at first glance, his appearance there (together with that of the Iranian president, Mahmoud Ahmadinejad) made sense, given Venezuela's growing engagement in extrahemispheric affairs and its policy of reinforcing its position as a supplier of petroleum. Chávez also had secondary motives for the increased intensity, visibility, and pace of his international diplomatic and economic activities. In 2006 alone, he made state visits to Belarus, Russia, Qatar, Iran, Vietnam, Mali, Benin, China, Malaysia, Syria, and Angola. In early 2009, he concentrated on shoring up his relationship with Venezuela's allies in the Alternativa Bolivariana de las Américas (ALBA): Cuba, Bolivia, Ecuador, Nicaragua, Dominica, Antigua/Barbuda, St. Vincent, and the Grenadines.

Most of his pronouncements in these countries featured the overwrought anti-American rhetoric for which he is famous, as well as appeals for fraternal cooperation to limit the hegemonic or unilateral influence of the United States. These initiatives have not always been successful. In 2006, for example, Venezuela made a concerted (and costly) effort to gain a seat on the United Nations Security Council. After a prolonged struggle and numerous ballots in the General Assembly, the United States rallied enough votes to block the Chávez initiative.

Quite evidently, Chávez has not limited his foreign initiatives to the Western Hemisphere. As Ralph Clem explains in his chapter in this volume, the strengthening of Venezuela's ties with Russia, in particular, serves both countries well. In September 2008, Chávez made his sixth visit to Russia, where he met with President Dmitry Medvedev and his predecessor and now prime minister, Vladimir Putin. The two sides announced a substantial new arms purchase ($2 billion), bringing total Russian arms sales to Venezuela over the last several years to around $6 billion, as well as an expanded oil and natural gas consortium. Russian energy firms are already operating in Venezuela. The meeting was also an opportunity to engage in some anti-U.S. rhetoric, an activity that has become increasingly popular in Russia of late, and in which Chávez has proved to be a master.[25] The Russian Air Force deployed two long-range bombers to Venezuela in August 2008, the Russian Navy sent a squadron of warships in late November 2008 for joint exercises in the Caribbean, and Russian president Medvedev visited Caracas in November 2008 as part of a Latin America tour. Although none of these moves has generated much official concern from the U.S. government, they are symbolic of the "rattle-the-cage" approach that Russia has adopted recently vis-à-vis the United States, perhaps to demonstrate its displeasure over (now abandoned) plans to install U.S. missile defense systems in Poland and the Czech Republic, and for which reason Russia's friendship with Chávez permits a convenient tit for tat.

Venezuela's recent foreign trade deals connect directly to both the paramount energy security issue and the desire to limit U.S. influence, but they also illustrate the limitations inherent in the inertial international economic system. Venezuela's 2008 oil deal with China is an example; building on a 2004 agreement, the revised pact raises the target for providing Venezuelan crude to help satisfy China's insatiable demand for energy, and envisions joint refinery construction projects.[26] For Venezuela, entrée into the Chinese market allows for at least a modicum of diversification away from the United States as the principal destination for Venezuelan oil. Because of the realities of distance and transportation costs, however, and the need for special refineries to process the heavy Venezuelan crude, China's reliance on Middle Eastern oil will actually increase over time.

Returning to our opening point about Venezuela's long-term reliance on oil revenues as the principal driver behind its foreign policy, we must recall that this one resource provides 90 percent of the country's export earnings, 50 percent of its federal budget, and 30 percent of GDP.[27] Further, the most recent data (2008) reveal that the United States remains by far the principal destination for Venezuelan exports (mainly crude oil, but not exclusively) and provides more than one-quarter of it total imports, despite the political estrangement of the two countries.

Just as the dramatic rise in oil prices provided the basis for rapid growth in Venezuelan GDP after 2004, in late 2008 and early 2009 the equally stunning decline in commodity prices and runaway inflation brought about by economic recession generally jeopardized Venezuela's ambitious domestic social and economic development plans, as well as its Bolivarian foreign policy.[28]

There is ample evidence that Venezuela is having to cut back on its ambitious petro-politics. Simón Romero of the *New York Times* reports that in 2009 Venezuela has announced plans to spend $6 billion abroad, compared to $79 billion in 2008.[29] Similarly, the triad of Russia, Iran, and Venezuela, half-jokingly referred to as the "Axis of Diesel," was considerably less well-positioned to pursue its agenda of oil, arms, and ideology, and its ambitious plans for forging international relations of a new type may be much longer in gestation.[30] In this volume, Jorge Castañeda argues that Chávez has few reasons to be sanguine about the success of his "Bolivarian" designs, given the new global and hemispheric geopolitical realities.

That fact is reflected in the critical slant of most of the essays in this volume. And yet, not everyone shares this oppositional stance. As the chapter by Norman Girvan demonstrates, Chávez initiatives such as Petro-Caribe and ALBA enjoy real popularity in the Greater Caribbean. Anthony Maingot, while not disagreeing with Girvan, explains the reasons why certain Caribbean countries have resisted joining Petro-Caribe. The situation is not much different in Europe: Julia Buxton describes the very favorable reception given the "Bolivarian social agenda" by many on the European left.

The usually skeptical Fernando Coronil believes that during the Chávez presidency the people have become "*el soberano*" (sovereign). It would take a lot, he argues, to turn this principle of sovereign consent into a fully democratic reality, but he considers Chávez's leveling efforts "a formidable accomplishment."[31] Indeed, in the 2008 Latinobarómetro poll, Venezuelans were second only to Uruguayans in expressing "satisfaction with the way democracy works in your country."[32] More than a third of Venezuelan respondents said inequalities had diminished during the Chávez years, a possible explanation for "the popularity of Mr. Chávez, an oil rich strongman." This is not the whole story, of course; the poll's editors went on to explain that, in Venezuela, support for democracy "may have been boosted this year among opponents of President Hugo Chávez, after their victory in a referendum on constitutional change last December." If we accept that interpretation, then by logical extension the 65.5 percent of the electorate who voted in the November 2008 municipal and state elections further strengthened support for democracy. Although Chávez's United Socialist Party of Venezuela (PSUV) won seventeen of twenty-two governorships, the opposition coalition triumphed in the five most important states in terms of population, commerce, and oil, as well as in the all-important mayoralty of Caracas. The country's three most influential newspapers proclaimed the results a victory for democracy and the beginning of a new phase in Venezuelan politics.[33]

The unanswered question is whether this divided coalition of parties represents not just an opposition but an alternative to the regime. The findings of a recent comprehensive poll reveal the contradictions and the complexity of the situation:[34] 56 percent of the population believes the country is on the wrong path, and only 15 percent believe that it will improve in 2009–10. Only 9 percent believe that Chávez's "Socialism of the Twenty-First Century" is a promising option, especially if it leads to a Cuba-like situation. By large majorities, poll respondents reject the nationalization of private property and the ongoing centralization of power in the presidency. Only 3 percent believe that the country's problems lie with "the oligarchy" or "imperialism," both central themes of President Chávez's rhetoric. And yet, 43 percent say they would vote for Chávez, versus 39 percent for any candidate of the present opposition. On the other hand, if offered a "new" opposition candidate, 49 percent would vote for him or her, versus 40 percent for Chávez. The challenge for the opposition is to find a candidate who can articulate a program that can appeal to such a highly politicized electorate. Beyond the issue of political appeal, in the final analysis much will depend on the international price of oil. *Caudillo*-led populist parties, such as Chávez's PSUV, must be constantly nurtured by a flow of funds. With no hard-and-fast structure of succession, a *caudillo* without money soon loses legitimacy and democratic appeal. The alternative to democracy, *caudillismo*, of course, made Bolívar despair, but it is precisely what the philosopher of dictatorship, Vallenilla Lanz, recommended for Venezuela: rule by a "democratic Caesar." As difficult as it is to predict any such possibility, those interested in the future of Venezuela should keep an eye on two factors: the personal whims of President Hugo Chávez and the international market and price of oil. The chapters in this book are intended to help sharpen that focus, all the while admitting that a final judgment on the Venezuelan situation is not possible as of this writing. That said, and as we have already noted, certain secular and systemic trends are evident in the nation's economy, politics, and foreign policy. On July 1, 2009, the three major Venezuelan newspapers carried as a special insert a "document" from both the National Academy of Economics and the Academy of Political and Social Sciences.[35] The general thrust of the declaration was that the government's policies were leading the country in the wrong direction. The document cited three basic problems:

1. The decaying infrastructure of the country, from oil refineries to roads, water, and electrical networks;
2. The "moral decomposition through rampant corruption" among the upper levels of the regime; and
3. The inefficiency of the ever-increasing centralization of political and economic power.

Events occurring only five months later appear to confirm the fears of the two academies. A massive scandal among several banks owned by men close to the regime forced President Chávez to close these banks and arrest those executives who did not manage to flee the country. The deterioration of the infrastructure was evident in the drastic rationing ordered for electricity and water consumption in several major cities.

In 2009, Venezuela's neighbors were rapidly recovering from the global recession, but Venezuela was not. In fact, it had entered into recession. The effects of a reduction in oil revenues were evident. Oil export revenues in 2009 were half of what they were in 2008 which explains why the economy grew by 4.8 percent in 2008, but contracted by 2.4 percent in the second quarter of 2009, and why Venezuela's trade portfolio for the third quarter of 2009 was $5.1 billion less than the $17.3 billion surplus in 2008.[36]

Arguably, it is in the area of foreign policy, however, that Chávez seems to have chosen to stake out a highly perilous course. By the middle of 2009 Chávez had returned to his extra-continental friends, and had radicalized his rhetoric. Once again in Russia, for the ninth time, he praised Lenin as one of that country's greatest sons, joining Nicaragua as only the second country to extend diplomatic recognition to Georgia's breakaway enclaves of Abkhazia and South Ossetia, and purchasing (on Russian credit) an additional $4.4 billion of Russian military hardware.[37] In Belarus, he proposed to President Aleksandr Lukashenka a "union of free republics." In Libya, he declared that Jesus Christ and the Prophet Mohammed, like Bolívar, were "true socialist fighters." It was in Iran, however, that he solidified and promoted his most risky initiatives. Chávez has always been an enthusiastic supporter of the efforts of President Mahmoud Ahmadinejad to turn Iran into a nuclear nation.[38] Chávez was the first to congratulate Iranian President Ahmadinejad on an electoral victory that was widely criticized as irregular by much of the world. The two countries had already negotiated or signed bilateral agreements in the areas of trade and banking, and were committed to various other forms of cooperation (e.g., in the areas of science and technology, cultural exchanges, and agrarian development) when, in mid-2009, Chávez announced that Venezuela and Iran would construct a "nuclear village" in Venezuela and that Iran would help develop Venezuela's uranium deposits. This admission came just as the international community was condemning the existence in Iran of an erstwhile secret site for the enrichment of weapons-grade uranium and had begun considerations of implementing stricter sanctions on that country. All of this was revealed by the *New York Times,* which editorialized that Iran was "The Big Cheat."[39]

As this is being written, there is much speculation as to what motivates Chávez to engage so forcefully with countries that are clearly in the eye of several geopolitical storms. In a stinging speech, Manhattan's highly respected district attorney, Robert Morgenthau, alleged that both Chávez and Ahmadinejad have been cooperating with Hezbollah in collecting money through the drug trade, and that Chávez's motivation was that he was "bent on becoming a regional power."[40] Perhaps most devastating, however, was the opinion expressed by London periodical, *The Economist,* which noted that Chávez had used his most recent tour to cement "an anti-American alliance with Iran, Syria, Belarus and Russia," and which then concluded that "most of the regimes he is cultivating in this enterprise are marked by rigged elections, media censorship, the criminalization of dissent and leaders for life." The journal then asked, "Is this the future of Venezuela?"[41]

While all this sums up the situation as of early 2010, and with polls all showing an erosion in public support for the regime, it would be a serious mistake to underestimate Chávez's skills at populist recoveries. Just like he turned the prolonged strike at PDVSA to his advantage, he has now taken on the mantle of corruption-fighter by closing seven banks. The fact that these banks were owned by members of his intimate circle (the so-called "Boligarchs," or *boliburguesía*) does not appear to have affected him negatively, since it also highlights his constant criticism of oligarchs and capitalists. Beyond strategy, and the fact that the opposition seems incapable of presenting a united front, President Chávez still has access to an ample campaign fund. As of January 2010, it is calculated that President Chávez holds a total of US$41.1 billion in various institutional accounts under his personal control and completely outside any legislative controls.[42] It is expected that government spending will increase considerably before the legislative elections for the National Assembly, scheduled for September 2010. Few are predicting an early demise of President Chávez's petro-politics and petro-diplomacy.

## Notes

1. Fernando Coronil, *The Magical State: Nature, Money, and Modernity in Venezuela* (Chicago: University of Chicago Press, 1997), 20.

2. "Contestación de un americano meridional [Bolívar] a un caballero [Henry Cullen] de esta isla [Jamaica]," Kingston, 6 de septiembre, 1815," in *Simón Bolívar—Doctrina del libertador*, ed. Manuel Pérez Vila (Caracas: Biblioteca Ayacucho, 1976), 55–75 (quotation on p. 74).

3. John Lynch, *Simón Bolívar, A Life* (New Haven, Conn.: Yale University Press, 2006), 92.

4. Pérez Vila, *Simón Bolívar—Doctrina del libertador*, 327.

5. See Edwin Lieuwen, *Venezuela* (London: Oxford University Press, 1961), 168.

6. Editorial, *New York Times*, December 15, 1908, cited in Rómulo Betancourt, *Venezuela: Política y petróleo* (Mexico City: Fondo de Cultura Económica, 1956), 27.

7. Laureano Vallenilla Lanz, "Cesarismo democrático: Estudio sobre las bases sociológicas de la constitución efectiva de Venezuela" (Caracas: Tipografía Garrido, 1961 [1919]), 206–7.

8. Robert J. Alexander, *Rómulo Betancourt and the Transformation of Venezuela* (New Brunswick, N.J.: Transaction Books, 1982), 524.

9. Betancourt, *Venezuela*, 774.

10. Armando Rojas, *Los creadores de la diplomacia venezolana* (Caracas: Imprenta Nacional, 1977), 24.

11. Harold Trinkunas, "Energy Security: The Case of Venezuela," in *Energy Security and Global Politics*, ed. Daniel Moran and James Russell (New York: Routledge, 2008), 175–87.

12. Arthur M. Schlesinger Jr., *A Thousand Days: John F. Kennedy in the White House* (Boston: Houghton Mifflin, 1965), 766.

13. John Martz, "Venezuelan Foreign Policy toward Latin America," in *Contemporary Venezuela and Its Role in International Affairs*, ed. Robert Bond (New York: New York University Press, 1970), 158.

14. Ibid., 159.

15. Franklin Tugwell, "Venezuela's Oil Nationalization: The Politics of Aftermath," in Bond, *Contemporary Venezuela and Its Role in International Affairs*, 115.

16. Martz, "Venezuelan Foreign Policy toward Latin America," 160.

17. Simón Romero, "Venezuela Markets Itself as a Hub for Leftists," *International Herald Tribune*, November 11, 2008, at www.iht.com.

18. Ricardo Hausmann, quoted in Fernando Coronil, "It's the Oil, Stupid!!!" *ReVista: Harvard Review of Latin America* (Fall 2008): 23.

19. "Country Briefing: Venezuela," *The Economist*, Intelligence Unit Report (November 26, 2008).

20. *El Universal* (Caracas), September 25, 2009, 1.

21. *El Tiempo* (Bógota), September 24, 2009, 1.

22. STRATFOR, "Colombia, Venezuela: Chávez Threatens Trade," Global Security and Intelligence Report, August 28, 2009, 3.

23. *Tal Cual* (Caracas), September 16, 2009, 1.

24. World Bank expert, cited in *El Nuevo Herald* (Miami), September 25, 2009, 1.

25. Luke Harding, "Venezuela: Hugo Chávez in Moscow to Sign $2bn Arms Deal," *The Guardian*, July 23, 2008; "Hugo Chávez' Rating Soared a Billion," *Kommersant* (Moscow), September 27, 2008.

26. Warren Bull, "Venezuela Signs Chinese Oil Deal," *BBC News/Americas*, September 25, 2008, at http://news.bbc.co.uk.

27. Central Intelligence Agency, *World Factbook-Venezuela*, at www.cia.gov; data are for 2007.

28. A statement signed by eighteen of Venezuela's most-respected academics speaks of erratic processes of "expropriations," "improvised" economic planning, and the favoring of "circumstantial geopolitical interest over national ones" (see "Ante la situación economica nacional," *Pensar en Venezuela*, May 12, 2009, at www.pensarenvenezuela.org.ve).

29. *New York Times*, May 20, 2009, 6.

30. "An Axis in Need of Oiling," *The Economist*, October 25, 2008, 71–72.

31. Fernando Coronil, "Chávez's Venezuela," *ReVista: Harvard Review of Latin America* (Fall 2008): 3.

32. "Democracy and the Downturn," *The Economist*, November 13, 2008.

33. See editorials in *Tal Cual, El Nacional,* and *El Universal*, November 24, 2008.

34. "Informe Final," Agencia Venezolana de Inteligencia Hinterlances, *Democracia Participativa,* June 27, 2009, at libertad@democraciaparticipativa.net. Very similar responses were given to a later poll, conducted by Alfredo Keller y Asociados (see "Las piedras en el camino de la revolución: Lo que piensa la opinión pública" [Caracas, July 22, 2009]).

35. "Venezuela ante la crisis: Documento que presentan a la opinión pública nacional la Académia Nacional de Cienciás Económicas y la Académia de Cienciás Políticas y Sociales" (Caracas, July 1, 2009).

36. *Financial Times,* November 19, 2009, at www.ft.com.

37. *New York Times,* September 11, 2009, A-11.

38. See "La presencia de Teheran aumenta en Latinoamerica," *El Nuevo Herald,* November 24, 2008, 6B.

39. *New York Times,* September 26, 2009.

40. Robert M. Morgenthau, "The Emerging Axis of Iran and Venezuela," *Wall Street Journal,* September 8, 2009. Morgenthau's remarks were originally delivered as a speech to the Brookings Institution on September 7. The Venezuelan ambassador to the United States responded to Morgenthau in a scathing letter in which he refuted Morgenthau's charges point by point. The letter was circulated on the Internet on September 11.

41. *The Economist,* September 22, 2009, www.economist.com.

42. *El Nacional,* December 27, 2009, 1.

# 1

# The Logic of Venezuelan Foreign Policy during the Chávez Period

HAROLD A. TRINKUNAS

To some casual outside observers, Venezuela's foreign policy under the Chávez administration appears radical and increasingly militarized. The war scare with Colombia that followed the Colombian incursion into Ecuador to target FARC leaders generated headlines in part because of the mobilization of the Venezuelan armed forces. Regional leaders have pointed to the large amounts of money that Venezuela is spending on arms purchases, in which it now competes for top place with Chile and Brazil. There is also concern in some quarters over the ideological component of Venezuela's foreign policy. The Bush administration denounced Chávez's close ties to Iran and expressed caution over his growing military and economic ties with Russia and China. Colombia has complained repeatedly about Chávez's sympathy for the FARC, and governments in Peru, El Salvador, and Nicaragua (before the election of Daniel Ortega) have reported Bolivarian propagandizing by Venezuelan-backed civil society groups.[1]

At least in tone, Venezuela's foreign policy appears truly revolutionary and overtly ideological by comparison with that of previous administrations. The verbal pyrotechnics that characterize Chávez's interactions with some foreign leaders, including George W. Bush, the Spanish monarch, and the German chancellor, contribute to this impression. If we examine the components of Venezuela's foreign policy strategy, however, we see elements of continuity with the past. War scares with Colombia are nothing new, and arms spending sprees characterized the administrations of General Marcos Pérez Jiménez (1952–58) and Carlos Andrés Pérez (1974–79), both of whom

governed during economic boom times. Even regime export has been a traditional feature of Venezuelan foreign policy, with Pérez Jiménez supporting
fellow dictators Somoza and Trujillo in the 1950s and Betancourt aiding democratizers across the Caribbean at the beginning of Venezuela's democratic
period in the 1960s.[2]

Venezuela's foreign policy today seeks to defend the Bolivarian revolution;
promote a sovereign, autonomous leadership role for Venezuela in Latin
America; oppose globalization and neoliberal economic policies; and work
toward the emergence of a multipolar world with checks on U.S. hegemony.[3]
The Chávez administration sees the United States as the main threat to the
regime's survival, a suspicion reinforced by the U.S. role in the 2002 coup attempt against him and support for his regional rival, President Álvaro Uribe
of Colombia.[4] Venezuela's rhetoric during the recent Colombia-Ecuador crisis reflects this perspective.[5]

Venezuela's foreign policy is revolutionary not for its methods but for
its objectives. It is driven by a profound departure in its leader's strategic
analysis of the international order. For most of the twentieth century, Venezuelan leaders believed that they benefited from a friendly relationship with
the United States, and, given their shared interest in oil, they occasionally
sought to establish a special relationship between the two countries. What is
new about foreign policy thinking in the Chávez regime is its identification
of the United States as the main threat to Venezuela, and of U.S. hegemony
as a threat to the international community that should be contained by the
development of alternative poles of global power.

The discussion that follows examines the logic of Venezuela's foreign policy under the Chávez administration and assesses its prospects for success or
failure, drawing upon the existing literature on the foreign policy behavior
of states in Latin America to provide a context for understanding Venezuela's
international behavior. It shows how Venezuela's behavior has been shaped
by these constraints, exploring both realist and identity-based explanations
for what we have observed during the Chávez administration. A concluding
section examines the implications for U.S. policy in Latin America.

## Theoretical Approaches to Explaining Venezuela's Foreign Policy

Three aspects of Venezuela's contemporary strategic behavior have caused
alarm among outside observers: the pursuit of a major weapons acquisitions

program with Russia; the formation of alliances with radical, authoritarian, or less cooperative states in the international system and the sponsorship of like-minded regimes in Latin America; and the use of oil wealth as an instrument of foreign policy. None of these behaviors is uncommon in the history of Venezuela or Latin America, and its motivations are clear once we take seriously Chávez's view of the United States as the main threat to the Bolivarian regime.[6]

Several theoretical arguments help explain Venezuela's foreign policy behavior. Among the central tenets of realist and neorealist thinking about international relations is the notion that an anarchic international system leads states to look to their own security. The typical response is self-help behavior, either through strengthening military and other capabilities or fostering alliances to balance threats.[7] Stephen Walt argues that states pay attention not only to relative capabilities when making calculations of threat, but also to the identity and nature of other powers and their intentions. For a medium power such as Venezuela, threat perceptions matter greatly. For much of the twentieth century, Venezuela viewed relations with the United States as positive and non-threatening, and tended to either bandwagon with or show sympathetic neutrality toward the regional hegemon. With worsening U.S.-Venezuela relations, we can expect to see a shift toward balancing behavior.

Another school of thought about international relations in the developing world, so-called "peripheral realism," suggests that anarchy is not really an accurate description of the international system. It argues that states play different roles, as order-givers (developed core states), order-takers (peripheral states that accept the existing international order), and order-breakers (peripheral states that seek to change the international status quo). Order-breaking countries tend to have fraught and conflictive relations with order-makers, such as the United States. At various points in history, some Latin American states have migrated to the order-breaker category: Cuba in 1959, Nicaragua in 1979, and Argentina from 1976 to 1987 (at least in terms of missile proliferation).[8] Venezuela's long-range international objectives, especially its attempts at regime export, allow us to add it to this group.

One explanation for regime export behavior focuses on identities. An actor's identity is its sense of who it is and who it is not, what it stands for and what it is against.[9] Social and political identities can be defined primarily in terms of nationality, but in Latin America, as in other historical and geographic contexts, ideological distinctions or alternative political projects are also important. The Bolivarian revolution has no rigid ideology or system of government to export, but rather a loose set of ideas, attitudes, and predispositions that cluster around a populist philosophy of government and a rejection of free markets, globalization, and U.S. hegemony.

Finally, we should keep in mind that domestic concerns are an important component of Venezuela's foreign policy. Stephen David's concept of "omnibalancing" helps identify the security priorities of leaders in the developing world by considering distinctive features of the environment they confront: uncertain sovereignty, weak state structures, competing domestic power interests, an unsure state monopoly of the instruments of force, and disloyal oppositions.[10] Threatened simultaneously by internal and external enemies, leaders must balance in all directions at once. Given the limited conventional capabilities of most Third World states, internal threats can provide opportunities for outside actors to wield inexpensive and effective influence over the policies of unstable neighbors. The relative weakness of Latin American states may give a leader such as Chávez an opportunity to export his preferred regime type, but it also makes him vulnerable to having some other regime type foisted on him. This dynamic helps explain Chávez's trenchant relationship with domestic opponents and the "sovereignty-hardening" measures he has pursued to sever their relations with foreign partners.

## The Logic of Venezuela's Foreign Policy

Venezuela's international policies seek minimum and maximum objectives: defense of the revolution at home and the reordering of the international system into a multipolar world. Some of Venezuela's foreign policy actions reflect realist concerns over the imbalance in military power between the U.S. and Venezuela and the growing military power of Colombia. This preoccupation drives the Chávez administration's efforts to engage in self-help by increasing its military potential and adopting new military doctrines for surviving an asymmetric confrontation with an external power. Identity, however, also plays a major role in the choice of allies, as Chávez seeks out partners who are ideologically in sympathy with his Bolivarian revolution.[11] Maintaining and expanding this coalition of like-minded governments, parties, social movements, and activists through the use of Venezuela's oil power is both a regime defense strategy and an effort to reorder the international community. Three aspects of Venezuelan foreign and defense policy are especially important to these objectives: increasing Venezuela's relative military power, building a network of new alliances, and using the country's growing oil wealth.

## Increasing Military Capabilities

Venezuela is only the latest Latin American state to take on an "order-breaker" role, and given that the post–Cold War period is generally thought of as a period of U.S. hegemony, it is not surprising that Venezuela would perceive the U.S. as a significant threat. Ideology and past experience also influence Chávez's view of the United States as an offensively minded power. With self-help and alliance-formation the expected impulses of states confronted with a threat, particularly one of the dimensions of the threat posed by U.S. capabilities, we can expect to see Venezuela increase its arms acquisitions, reform the military, expand reserve forces, and step up military exercises.

Under Chávez, Venezuela has become one of Latin America's largest arms importers, on a per capita expenditure basis. Imports include the purchases of Sukhoi fighter aircraft, Ilyushin transport aircraft, assault and transport helicopters, diesel submarines, and small arms from Russia, as well as long-range radar and jet training aircraft from China.[12] Some of these acquisitions represent renewal of older equipment, such as the Mirage and F-16 aircraft, and others new capabilities, such as long-range radar and air defense missile systems. The growing strength of Colombia, a close U.S. ally, is part of the rationale for Venezuelan arms expenditures, although this is of course not its only justification for acquiring new weapons systems: another is the threat posed by the United States itself.

Venezuela has engaged in a prolonged internal discussion within its armed forces over appropriate military doctrine given its strategic analysis of the United States and Colombia as major threats. The result is a new doctrine, Defensa Integral de la Nación (Integrated National Defense), which envisions prolonged popular resistance to an external invasion. Venezuelan leaders have spoken positively of Vietnam's resistance to the United States in the 1960s, Saddam Hussein's firing of Kuwaiti oil fields in 1990, and Iraq's present insurgency as examples of this strategy. To break the traditional doctrinal links between the U.S. and Venezuelan militaries, Chávez closed down the Military Group attached to the U.S. Embassy in Caracas, the entity that would normally handle all security assistance and cooperation exchanges between the two countries. He has made the expansion of the reserves to over a million troops the cornerstone of his new defense policy and proposed the creation of a millions-strong territorial guard.[13] Weekend training of reservists has begun; the professional military education provided in Venezuelan war colleges now includes a greater focus on asymmetric warfare; and the military has held numerous conferences on "fourth-generation warfare."[14] War games and exercises simulating popular resistance to a U.S. invasion have also been staged in recent years, although reporting suggests that they are not taken very seriously by the civilian participants.[15] The effort that has been expended to produce changes in doctrine, however, suggests a serious commitment at the senior political and military leadership levels.

## Alliance Formation: Ideology, Identity, and Balancing Behavior

Conservative analysts in the United States have reacted with particular concern to Venezuela's efforts to export the Bolivarian revolution across Latin America, as attempted by Cuba in the 1960s and Nicaragua in the 1980s. Attempts at regime export are not uncommon in Latin America; in addition to the previous two examples, Caribbean democrats and dictators engaged in subversion and countersubversion in each other's countries during the 1940s and 1950s, with democratizers organizing the Caribbean Legion and the dictators clubbing together as the Internacional de las Espadas.[16] The efforts of the Contadora Group to promote democratization in Central America as an instrument for conflict resolution fall into the same category.[17] What explains this behavior, and to what extent is Venezuela pursuing it?

State leaders, Chávez included, identify other actors as potential adversaries or allies in terms of their perceived identities—that is, based on their perceptions of the actor's "inherent" character, nature, or ideology. If we return to Walt's analysis of the components of threats, actors' identities serve as fundamental criteria for shaping assessments of offensive (or benign) intent. Regime export can be part of strategies to replace actors whose identities label them as adversaries. This can be attempted through both overt and direct means (e.g., military force through conventional war) and, more realistically given the relatively weak military capabilities of most Latin American states, covert and indirect measures (funding, training, or organizing insurgents).

In the case of the Chávez administration, the pattern of support for like-minded states, political parties, NGOs, and other non-state actors is part of the process of developing a community of actors who share an ideological affinity, as well as changing the identities of regimes that may pose a threat to the Bolivarian revolution. Chávez's sympathy and support for the FARC, for example, can be understood as a way for him to influence the domestic politics of neighboring Colombia. His support for the FMLN and FSLN in Central America helps him secure influence over that region's politics. In the case of Peru, Chávez support for the Ollanta Humala campaign was an effort to prevent another Alan García presidency.

Chávez has had relative success in building strong relationships with other Latin American regimes that share his ideological affinities, particularly Bolivia, Ecuador, and Nicaragua. Economic interests and trade also play a role, of course. Venezuelan support has helped Argentina resist pressure from international financial institutions, and Argentine food exports on preferential terms have enabled the Chávez administration to respond to the chronic food shortages that are one of the main sources of popular discontent in Venezuela. Venezuela has invested heavily in providing development assistance to like-minded politicians in other countries, and the ALBA, Chávez's alternative to the FTAA, has gained a few adherents, mainly among states that receive economic assistance from Venezuela.

The support of like-minded states is particularly important in Latin America because of an unwritten post–Cold War code that prevents the United States from over-intervention in the region. Part of this understanding is limited U.S. interference in Latin American states that are democracies. Since the definition of democracy is not simply academic, but political, Chávez's allies help promote the idea within the region that Venezuela is fully democratic, regardless of what opposition groups say. This democratic identity helps shield Venezuela from overt U.S. meddling, so it is very important to the Chávez administration's strategy of defending the Bolivarian revolution.

Not all of Venezuela's alliances are based on ideology or identity. Some are grounded solidly in realist concerns over the balance of power. Given Chávez's desire to acquire new military capabilities and his perception of the United States as the main threat to his regime, he can be expected to purchase new systems from non-U.S. sources. In the highly interdependent international arms production industry, however, weapons integrate systems sourced from multiple providers. The most advanced guidance, targeting, navigation, and command and control systems are almost entirely licensed from the United States or incorporate U.S. technologies. To avoid the pro-liferation of such technologies to undesirable elements in the international system, the United States demands that licensees seek permission before transferring its technology to third parties. This mechanism allowed the U.S. to block arms sales by Spain and Brazil to Venezuela between 2004 and 2006. Venezuela's increasing reliance on Russia for weapons technology can be explained by the fact that Russia is one of the few major arms producers that typically does not incorporate U.S.-licensed technology in its military equipment, and is therefore able to transfer technology freely to states such as Venezuela that have been "blacklisted" by the United States.

Clearly, purely ideological explanations are not the reason for Venezuela's selection of allies such as Russia, Iran, China, Iraq (before 2003), Zimbabwe, and Libya (before Ghaddafi abandoned his nuclear program). Given Venezu-ela's role as an order-breaker in an international system in which the United States has long been an important order-maker, it is in Venezuela's interest to seek out alliances with other order-breakers, or at least "order-skeptics" such as Russia and China. The fact that these states are authoritarian, anti-demo-cratic, or simply disagreeable is not surprising, since the international order they are skeptical of tends to promote liberal formulations of representative democracy, human rights, free markets, and free trade. In the post–Cold War period, the international community has at times displayed a permissive attitude toward intervention when such principles are violated, justifying meddling in order-breaking countries on national security or international good governance grounds. This history contributes to the threat perceptions of the Venezuelan leadership, incidentally confirming Walt's hypothesis that states respond to perceptions of offensive intent with balancing rather than bandwagoning behavior.

Stephen David's view of the way leaders of weak states respond to threats via omnibalancing also helps explain Chávez's attitude toward his domestic opponents. Their connections to international democracy promotion and globalization movements and organizations make opposition actors con-duits for anti-Bolivarian ideas and (very rarely) checks on the regime's do-mestic policies. The Chávez administration takes a suspicious view of U.S. support for Venezuelan opposition groups, even support that is intended mainly for democracy promotion.[18] It has responded with a number of sovereignty-hardening measures, including restrictions on foreign fund-ing, tougher reporting requirements, tax audits, informal intimidation, and police surveillance, to discourage contacts between the opposition and outside sponsors.[19]

## Oil as an Instrument of Foreign Policy

Venezuela has historically used its status as a major supplier of oil to world markets to support its foreign policy, and its foreign policy as a tool to enhance the rents it extracts from international oil markets. Venezuela holds some of the largest oil reserves in the world—the largest if ultra-heavy crude deposits are included in the calculation. Its oil strategy during the twentieth century can be characterized by a certain suspicion of the fairness of markets, which led it to promote the creation of OPEC and the national oil company, Petróleos de Venezuela (PDVSA). Until the Chávez administration, however, Venezuelan leaders were generally confident in the ability of diplomacy and law, as distinct from the untrammeled activities of buyers and sellers, to provide secure access for Venezuela's production to oil markets.[20] Prior to 1998, Venezuela benefited from its alignment with the West—more specifically, the United States—insofar as it did not have major external security concerns over its access to world oil markets. The U.S. was (and remains today) the major consumer of Venezuela's oil production and the major source of Venezuela's imports. Rather than challenging the status quo, Venezuela's main concern during this period was to maximize the rents that could be extracted from this system.

Since coming to office, Chávez has not hesitated to deploy Venezuela's oil power as an instrument of foreign policy. He has worked diligently to improve the cohesion and market-setting power of OPEC and tried to convince the organization to take a more radical stance in international politics. As relations with the United States have worsened, Chávez has sought alternative destinations for Venezuelan oil exports. The heavy and sour nature of Venezuelan oil, however, makes it difficult to refine, and most of the refineries with the technology to process Venezuelan crude are in the United States.

Chávez has dramatically increased his control over the nationalized oil industry to ensure that he is able to direct the country's oil rents to support his policy preferences. Long known for its autonomy and technical excellence, PDVSA experienced increasingly tense relations with the Chávez administration during his first and second terms in office, culminating in a general strike in 2003 to protest government interference in the company's internal meritocracy. Chávez responded with the mass firing of mid-level and senior managers and engineers and began partially renationalizing PDVSA's joint operating ventures with international oil companies. These measures have not only increased state control, but have also helped mask a decline in oil production due to persistent undercapitalization.

Venezuela has invested a substantial portion of its oil rents in international assistance programs designed to support alliance-building projects. Chávez has donated oil or money to poor communities across the Americas, including the United States, and he has resurrected an old policy of providing oil at preferential rates to Caribbean and Central American countries. He has also undertaken new initiatives to foster energy integration in the region to strengthen the role of government in exploitation and commercialization. Perhaps the most important target of Venezuelan oil assistance is Cuba, which receives more than $1 billion in oil per year. In return, the Venezuelan government receives assistance from Cuban doctors, educators, military trainers, and intelligence officers. Chávez has also targeted oil support to like-minded political movements, such as the FMLN in El Salvador and the FSLN in Nicaragua.

The effects of these measures are magnified by the sharp increase in the price of Venezuelan crude, which rose from less than $10 per barrel when Chávez was first elected to more than $120 per barrel in 2008. This spike has boosted the income of the Venezuelan state, and, consequently, its spending. As a result of the expansion of the state's role in the economy and the role of oil rents in subsidizing imports, Venezuela is spending money as fast as it is making it. This is a common pattern among oil exporters during commodity booms; if past experience is any guide, it can be predicted to introduce severe economic distortions, degrade institutions, and ultimately undermine the utility of oil as an instrument of foreign policy as all available rents are funneled toward maintaining domestic support for the regime.[21]

## Success or Failure of Venezuela's Foreign Policy and Implications for U.S. Policy

Merely having a strategy with some logical internal consistency is no guarantee that it will succeed. Now that the Chávez administration has been in power for a decade, it is time to assess whether it is meeting its overarching international objectives: defense of the Bolivarian revolution and the achievement of a multipolar world in which Venezuela has a global leadership role.

The status of the first of these objectives is relatively easy to assess. The Bolivarian revolution is still with us, although it is showing signs of internal fatigue. Chávez's failure to pass constitutional reforms in December 2007 and the suspension of education reform efforts designed to embed the revolution's ideology in the national curriculum are both indicators that he may be reaching the limits of his domestic political power.[22] In addition, the constant generation of new projects to address social ills, ranging from the civil-military Plan Bolívar 2000 to the new PDVAL food program, suggests that Chávez may be having trouble meeting domestic priorities and institutionalizing state programs to alleviate poverty.[23]

In terms of defending the revolution from external threats, however, the Chávez administration has succeeded. There are no signs of any international or regional consensus for placing limits on Venezuela's foreign or domestic policies; if anything, Latin American responses during the Colombia-Ecuador crisis in 2008 showed an unwillingness to side with Colombia and the United States. Despite warnings by some outside observers that the democratic process is deteriorating, there has been no serious talk at the regional or international level of outside assistance or intervention to reverse this process.

The Chávez administration has also been relatively successful in strengthening its use of oil power as an instrument of Venezuelan foreign policy. Rents from rising crude prices have allowed it to build a network of alliances with Central American, Caribbean, and South American states, most notably Nicaragua, Ecuador, and Bolivia. Strengthening state control over the Venezuelan oil industry has also improved Chávez's ability to forge new economic ties with China, Iran, and Russia. Not all of these efforts have been successful, as the failure to ensure the election of like-minded allies in Peru and Mexico demonstrates, but Venezuela's network of allies is considerably wider at the end of the first decade of the Chávez period than at its beginning.

Ultimately, Venezuela's association with some of the less savory members of the international system, such as the FARC or President Ahmadinejad in Iran, may turn out to be double-edged alliances. The 2008 border incident between Colombia and Ecuador, which included the seizure of computer files allegedly linking FARC leaders with Venezuelan military and civilian officials, may turn out to be costly for Chávez. It may not be enough to lead to a regional consensus that Venezuela has crossed the line and should be contained, but it may provide the U.S. with an excuse to label Venezuela as a state sponsor of terrorism.[24] Venezuela's ties to Iran and Chávez's controversial remarks about the Israeli-Palestinian conflict have also attracted negative attention beyond the United States.[25] Given the rising price of oil, however, the consequences of designating Venezuela a state sponsor of terrorism are sufficiently unpredictable that any administration will move cautiously in making this decision.

The success of the diplomatic and economic aspects of Venezuela's international strategy throws into relief the relative weakness of its military strategy. Despite billions of dollars in new acquisitions, the Venezuelan military has yet to integrate these systems into an effective national defense structure. The new reserve component is still a largely hollow force. The vaunted Sukhoi fighter jets are not operational, and Venezuelan pilots have only just begun to train in them. The mobilization of the Venezuelan army to the Colombian border during the March 2008 crisis highlighted numerous weaknesses in the readiness and logistics of the armed forces. In addition, the helicopters, fighters, submarines, and tanks that Chávez seeks to acquire from Russia and China are not the right instruments to deter the United States. Investment in massive quantities of light arms, portable missile systems, IEDs, and secure communications, together with a fully trained and staffed national reserve, would be much more appropriate to the prolonged popular war strategy advocated by current Venezuelan military doctrine.

Finally, Venezuela has made very little progress toward the multipolar world Chávez envisions. In part, this is due to his inability to build a broad and successful Latin American regional integration effort that combines political, military, and economic elements. Venezuela has proposed programs in all of these areas, but only ALBA has gained some traction in the region. Even MERCOSUR, led by political allies Presidents da Silva and Kirchner, has not approved Venezuela's full membership due to Chávez's unpopularity with the Brazilian Congress.[26] Political and military integration—much more difficult projects to accomplish, as the NATO and E.U. experiences suggest—has met with skepticism or silence from regional leaders. This is no surprise, considering Chávez's unpopularity among their constituents, who rated him as unfavorably as they did President Bush in the 2006 Latinobarómetro poll. Without a strong push from domestic interests, there is little reason to expect politicians across the region to pursue these objectives, and without a Latin American pole to form part of a multipolar international system, the maximum objective of the Bolivarian foreign policy agenda will not be achieved.

## Implications for U.S. Policy in Latin America

The irony is that the United States may be facing a multipolar world even without Venezuela. The rise of China and India as major international powers, commodity boom-driven economies in the global South, the relative weakness of the U.S. currency vis-à-vis other major currencies, and the drain on prestige and military power posed by the Iraq war all suggest that, in terms of relative gains, U.S. hegemony is not quite as certain as it was in the wake of the Cold War. The absence of a strong Latin America policy also undermines the United States' ability to influence the success or failure of Venezuela's international strategy.

Venezuela's foreign policy, given its record, poses a weak national security threat to the United States. At best, it rises to the level of a national security irritation, insofar as Venezuela interferes with the promotion of U.S. foreign policy objectives for the region. The quality of democracy in Venezuela has deteriorated, but that is really not a threat to U.S. security; rather, it is a threat to those in Venezuela and around the world who view democratization as a good idea. The inability to secure the FTAA has very little to do with ALBA and much more to do with the inability of the U.S. and Brazil to develop a package of mutual concessions on industrial and agricultural policies. Venezuela's oil sales to China and other non-traditional consumers is not really relevant to the United States as a consumer, since it can purchase replacement energy more easily than Venezuela can find alternative customers for its difficult-to-refine crude. In fact, the increased cost of selling Venezuelan oil to non-traditional consumers could be seen as a positive good, since it reduces the amount of rents the Chávez administration derives from the international system. Nor is Chávez's arms purchasing spree particularly threatening, because the weapons he is purchasing are precisely those that U.S. Cold War systems are able to destroy most easily, and the Venezuelan military has not demonstrated the ability to integrate the new equipment effectively. A regional arms race might be blamed on his activities, but given that most of the region is experiencing economic growth fueled by commodity booms, rising defense budgets—and all other government budgets—are only to be expected.

Chávez's efforts at regime export could be of concern to the United States, but it could be argued that Latin America's general drift to the left has very little to do with Venezuelan assistance. In fact, the ALBA member states are among the poorest countries in the hemisphere, cases where Venezuelan aid makes little difference. Chávez's inability to translate his political gains in Bolivia, Nicaragua, and Ecuador into a cohesive continental project, even at the economic level, suggests that the United States should moderate its concerns over Venezuela's international policies.

If the U.S. is serious about countering the Boliviarian agenda in the Americas, two considerations stand out: economic development and democracy promotion. One of the main engines driving the collaboration between Venezuela and like-minded regimes is money. Venezuelan aid is typically delivered without the kind of formal conditionality attached to U.S. or international assistance.[27] The FTAA has not gained traction with most Latin American states, and the neoliberal free trade agenda that underpins it is not as attractive to the region as it once was. Were the United States to propose a serious alternative development agenda for the region, one that helped the many economies whose growth is based on booming commodity prices invest in sustainable long-term growth and poverty reduction, this might provide a compelling countervailing narrative to the Bolivarian agenda.

In addition, given that domestic institutions have been the most effective check on the Chávez administration's policies, the U.S. should think seriously about how to conduct democracy promotion in Venezuela and other countries that have engaged in "sovereignty-hardening" measures. Due to the loss of credibility of U.S. democracy-promotion efforts under the Bush administration, this would mean providing funding to IGOs and NGOs rather than acting directly through the NED or State Department. At the very least, it also means stopping the habit of referring to democracy promotion as part of the United States' national security agenda, and returning to a policy in which support for democratization is linked instead to an interest in international good governance and the common good.

The change of administration in Washington after the 2008 elections is likely to be the shortest route to rebuilding U.S. influence in Latin America and checking Venezuela's foreign policy ambitions. A new president will have a chance to restore credibility to an agenda based on development and democracy promotion. In addition, to the extent that a new administration changes regional and international perceptions of the threat posed by U.S. hegemony, it will undercut the rationale driving Venezuela's foreign policy.

## Notes

Any opinions expressed in this paper are those of the author. They do not represent the official views of the Department of Defense or the U.S. government. Any errors contained herein are the responsibility of the author.

1. Richard Lapper, "Living with Hugo: U.S. Policy toward Hugo Chávez's Venezuela," Council on Foreign Relations, CSR No. 20 (November 2006); Howard LaFranchi, "Chávez's Anti-U.S. Campaign," *Christian Science Monitor*, September 29, 2006, www.csmonitor.com/2006/0929/p01s01-woam.html; Andrew Downie, "A South American Arms Race?" *Time*, December 21, 2007, www.time.com/time/world/article/0,8599,1697776,00.html.

2. Harold Trinkunas, *Crafting Civilian Control of the Military in Venezuela: A Comparative Perspective* (Chapel Hill: N.C.: University of North Carolina Press, 2005).

3. Julia Buxton, "Venezuela's Contemporary Political Crisis in Historical Perspective," *Bulletin of Latin American Research* 24, no. 3 (2005): 339–40; Elsa Cardozo da Silva and Richard S. Hillman, "Venezuela: Petroleum, Democratization, and International Affairs," in *Latin American and Caribbean Foreign Policy*, ed. Frank O. Mora and Jeanne K. Hey (Lanham, Md.: Rowman & Littlefield, 2003), 158–60.

4. Harold Trinkunas, "Defining Venezuela's Bolivarian Revolution," *Military Review* (July–August 2005): 39–44.

5. "Colombia Says No US Base Planned on Border with Venezuela," www.iht.com/articles/ap/2008/05/15/news/Venezuela-Colombia-US.php. For a fully developed argument regarding U.S. and Colombian threats to Venezuela, see James Petras, "The US/Colombia Plot against Venezuela," *CounterPunch*, January 25, 2005, www.counterpunch.org/petras01252005.html.

6. Trinkunas, "Defining Venezuela's Bolivarian Revolution."

7. Stephen M. Walt, "Alliance Formation and the Balance of World Power," *International Security* 9, no. 4 (Spring 1985): 3–43.

8. Carlos Escudé, "An Introduction to Peripheral Realism and Its Implications for the Interstate System: Argentina and the Cóndor II Missile Project," in *International Relations Theory and the Third World*, ed. Stephanie G. Neuman (New York: St. Martin's Press, 1998), 55–76.

9. John M. Owen IV, "Transnational Liberalism and US Primacy," *International Security* 26, no. 3 (Winter 2001/02): 117–52. On how state identities shape national security behaviors, see Ronald L. Jepperson, Alexander Wendt, and Peter J. Katzenstein, "Norms, Identity, and Culture in National Security," in *The Culture of National Security: Norms and Identity in World Politics*, ed. Peter J. Katzenstein (New York: Columbia University Press, 1996), 33–73; and Paul Kowert and Jeffrey Legro, "Norms, Identity, and Their Limits: A Theoretical Reprise," ibid., 451–97.

10. Steven R. David, "Explaining Third World Alignment," *World Politics* 43 (January 1991): 233–56. See also Steven R. David, "The Primacy of Internal War," in *International Relations Theory and the Third World*, ed. Stephanie G. Neuman (New York: St. Martin's Press, 1998), 77–102.

11. Steve Ellner, "Toward a 'Multipolar World': Using Oil Diplomacy to Sever Venezuela's Dependence," *NACLA Report on the Americas* (September/October 2007): 17; Robert Collier, "Chávez's Anti-US Fervor Emerging Force among Nonaligned Nations," *San Francisco Chronicle*, September 21, 2006, A-1.

12. "Venezuela Military Buildup at a Glance," *International Herald Tribune*, May 16, 2008, www.iht.com/articles/ap/2008/05/16/news/Venezuela-Military-Glance.php.

13. "Luchar con las alpargatas bien puestas cuando se nos gasten las botas," interview with National Security and Defense Secretary General Alí Uzcategui, Portal ALBA-TCP, www.alternativabolivariana.org/modules.php?name=News&file=article&sid=923.

14. Jorge Serrano Torres, "El sistema de inteligencia venezolano y la guerra asimétrica," VoltaireNet.org, December 12, 2005, www.voltairenet.org/article132343.html.

15. Omar Piña, "Plan Colombia: How US Military Assistance Affects Regional Balances of Power," M.A. thesis, Naval Postgraduate School, 2004.

16. Charles D. Ameringer, *The Caribbean Legion: Patriots, Politicians, and Soldiers of Fortune* (University Park: Penn State Press, 1996).

17. Michael Barletta and Harold Trinkunas, "Regime Type and Regional Security in Latin America: Toward a 'Balance of Identity' Theory," in *Balance of Power: Theory and Practice in the 21st Century*, ed. T. V. Paul, James J. Wirtz, and Michel Fortmann (Stanford, Calif.: Stanford University Press, 2004), 334–59.

18. N. Scott Cole, "Hugo Chávez and President Bush's Credibility Gap: The Struggle against US Democracy Promotion," *International Political Science Review* 28, no.4 (2008): 493–507.

19. Javier Corrales and Michael Penfold, "Venezuela: Crowding Out the Opposition," *Journal of Democracy* 18, no.2 (April 2007): 100–33; Carl Gershman and Michael Allen, "The Assault on Democracy Assistance," *Journal of Democracy* 17, no. 2 (April 2006): 38.

20. Harold Trinkunas, "Crafting Civilian Control of the Armed Forces," *Journal of Interamerican Studies and World Affairs* 43, no. 3 (Fall 2000): 77–109.

21. Francisco Rodríguez, "An Empty Revolution: The Unfulfilled Promises of Hugo Chávez," *Foreign Affairs*, March/April 2008, 49–62.

22. Elisabeth Young-Bruehl, "Behind the Student Movement's Victory," *The Nation*, December 6, 2007, www.thenation.com/doc/20071224/young-bruehl; Juan Forero, "In Venezuelan Schools, Creating a 'New Man,'" *Washington Post*, May 19, 2008, A19.

23. Rory Carroll, "The Long Slide," *The Guardian*, May 17, 2008, http://www.guardian.co.uk/world/2008/may/17/venezuela.hugo.chavez.

24. Juan Forero, "FARC Computer Files Are Authentic, Interpol Finds," *Washington Post*, May 16, 2008, A08.

25. Ellner, "Toward a 'Multipolar World.'"

26. "A Turning Point?" *The Economist*, June 5, 2007, www.economist.com/world/la/displaystory.cfm?story_id=9444187.

27. Moisés Naim, "Rogue Aid," *Foreign Policy*, March/April 2007, 95–96.

**2**

# Conflicting Goals in Venezuela's Foreign Policy

JAVIER CORRALES

The United States is again nervous about Latin America. The consolidation of Venezuela's Hugo Chávez, the most vehement U.S.-basher in the region since the Cold War, has caught the United States unprepared. Chávez issues threats against the United States almost daily, and the United States has begun to wonder whether any of these threats could materialize. Chávez has forced Washington to begin to draft all kinds of contingency plans, even "war" scenarios, in the Americas. These scenarios include Venezuela provoking a war against neighboring Colombia; spreading weapons and destabilizing democracies abroad; disrupting oil sales to the United States; providing financial support to Hezbollah, Al-Qaeda, or other fundamentalist movements; observing lax border, passport, and drug controls; and even acquiring nuclear weapons.

These are legitimate worries, but they miss a larger issue. The real challenge that Venezuela poses to the United States has less to do with aggressive actions that Venezuela could take against the United States than with something else in Venezuela's arsenal: a foreign policy based on generous handouts peppered with a pro-poor, distributionist discourse. For the purposes of this discussion, we will call this weapon "social power." In the United States we are used to discussing the requirements of "hard power" (military and economic might), and even "soft power" (the intrinsic appeal of values and institutions).[1] We spend less time discussing the requirements of social power—either as something to project or something to contain.

As a foreign policy tool, social power is a spectacularly effective way for world leaders to earn allies, and even admirers, abroad. Spending lavishly on social projects is a policy almost impossible to criticize. At a minimum, it serves to deflect potential criticism and scrutiny. Social power makes it essentially impossible to launch any type of multilateral initiative to contain the regime. Furthermore, social power is easy to emulate. Other regimes—with nastier, gutsier, and more competent leaders than Chávez—could replicate Venezuela's social-power foreign policy model and improve on it. The result could be the rise of meaner rogue states masquerading as international humanitarians. For all its power, the United States is simply unprepared to meet this potential new development in international politics.

## Social Power as a Foreign Policy Tool

Venezuela's foreign policy under Chávez has undergone two major changes relative to previous administrations. First, Venezuela has declared the United States the country's main adversary, and in response has adopted a policy of "soft balancing." Soft balancing is the term used to describe efforts short of military action to frustrate the foreign policy objective of larger nations.[2] Venezuela is, rhetorically at least, actively trying to counter and frustrate U.S. goals in the region.[3] Second, Venezuela has declared an overt commitment to promote development and, especially, help the poor at home and abroad. To further these goals, the Chávez government has gone on an international spending spree. It has offered investment to as many nations as possible, most of it billed as development aid. Venezuela's main innovation in foreign policy is to use this type of foreign economic largesse as a way to balance the United States. Few nations since the end of the Cold War have exploited this foreign policy tool to the same degree as Venezuela.

Venezuela's investments abroad have two salient characteristics. First, they are mostly carried out by the Venezuelan state (rather than private firms, as is typically the case with most FDI). Second, they include large sums for development projects. Gustavo Coronel estimates that Chávez has committed a total of US$43 billion in investments abroad. Of this total, perhaps US$17 billion (40.1 percent of the total) could be classified as social investments. This includes oil subsidies to Cuba and the members of PetroCaribe; the acquisition of Argentine commercial paper, which exempts the government from having to pay the IMF; cash donations to Bolivia; medical equipment for Nicaragua; even heating oil subsidies for more than one million U.S. consumers. Some estimates suggest that Chávez has provided or promised as much aid to Latin American countries, in real terms, as the U.S. spent on the Marshall Plan in Europe after World War II.[4] PetroCaribe alone, which represents an annual subsidy of US$1.7 billion, puts Venezuelan aid on par with OECD countries such as Australia, Belgium, Denmark, Norway, Portugal, Spain, and Switzerland.[5]

Every treaty Chávez signs seems to include an obligatory mention of development goals. ALBA, the Bolivarian Alternative for the Americas, is promoted as a "socially oriented" trade bloc dedicated to eradicating poverty. Multi-million-dollar monopoly investment deals with China and Iran come with special funds to promote "development," or to create "development banks." Foreign aid to Bolivia comes with real checks, issued by the Venezuelan embassy, for mayors to spend on "building hospitals." At OPEC meetings, Venezuela rebukes the Saudis for "not doing enough" to help the poor. In the United States, Venezuela has placed TV ads for CITGO, a company owned by the Venezuelan state, flaunting the fact that 228,000 households in the United States receive subsidized heating oil as a "gift of the people of Venezuela." PetroCaribe is a Venezuelan initiative to provide several small Caribbean countries with under 200,000 barrels a day of oil and petroleum products at preferential payment rates, with the savings to be used for development projects.[6] Jamaica has already used PetroCaribe loans for housing and other infrastructure projects. When a Brazilian plastics factory was shuttered in 2003 by its indebted owners, Chávez offered displaced workers subsidized raw materials in exchange for the technology to produce plastic homes in Venezuela. In 2007, Venezuela created a $20 million development assistance fund for Haiti, the poorest country in the Americas, aimed at investing in education, health care, housing, and other basic necessities.

Projecting social power as a diplomatic tool is not a Venezuelan invention. Great powers have used it. Small powers like Cuba have used it. Even previous Venezuelan leaders, such as Carlos Andrés Pérez in the 1970s, have used it. The Chávez innovation is to make social power the centerpiece of foreign policy, and to abandon the goal of promoting democracy abroad.[7] Few other countries have used social power to the same degree.

Compare Cuba's foreign policy during the Cold War and Venezuela's foreign policy during this oil boom. Both had a strong social policy component in their foreign policy toolkits. Cuba has been exporting doctors since the 1960s, but its most prominent exports during its heyday were guerrillas, weapons, and insurgency training.[8] The promotion of revolution abroad was always a far more important goal for Cuba than the export of doctors, by its leaders' admission. In Ché's words, Cuba's top interest was to create "one, two, many Vietnams." More than 350,000 Cuban personnel traveled abroad to plant guerrillas, assist existing ones, support socialist dictatorships, or conduct the actual fighting on their behalf.

Venezuela has adapted the Cuban model with a twist: let's create "one, two, many . . . clinics." If a foreign government accepts, what follows is more than just clinics. The Venezuelan government begins to extend its influence throughout the host country via direct cash payments with very few conditions other than: don't criticize me. Beyond that, recipients are free to use the money as they see fit.

Venezuela has thus developed a new export model. It is not so much the export of civil war, as in Cold War–era Cuba. It is not so much the export of weapons (as in both Communist and present-day Russia). It is certainly not the export of capital and technological know-how (as in the OECD countries) or the export of inexpensive manufactures (as in China). It's the export of corruption, or, more precisely, blank checks for governments and officials competing in democratic elections. The blank checks are billed as payments for social services, but in fact they represent unaccountable financing of corruption, campaigns, political movements, and governments.

And it works. As a publicity stunt, converting social policy into a primary foreign policy tool has brought Venezuela huge rewards. Domestically, it has allowed Chávez to win elections by seducing sectors on the left that are obsessed with international solidarity. Internationally, the effects have been more dramatic. Social power has brought Chávez two types of allies: other states that refuse to criticize him, especially if they receive petro-cash; and intellectuals on the left, especially in Europe, who admire almost any form of foreign aid.[9]

In Latin American diplomatic circles, Chávez's social-power foreign policy has essentially given him a shield against criticism. Governments on the left know better than to think of Chávez as a Venezuelan Mother Teresa. Everyone in the region understands that this is mostly a publicity stunt meant to camouflage serious domestic abuses and dubious international pretentions. Yet even these governments refuse to engage in a public fight with someone who gives the impression of having his heart in the right place. It's not just that center-left governments hope to receive special deals from Chávez (they do), it's also that they fear that picking a fight with Chávez will make him side with their most radical left-wingers at home, potentially destabilizing their governments.

In short, Chávez's social-power foreign policy has given him an impressive shield against international criticism, even by those who know better, and a reputation for humanitarianism among those who are less informed. This is an amazing foreign policy accomplishment. Undemocratic rulers worldwide, take notice: social power can save you from tough criticism of domestic performance. The Venezuelan foreign policy model holds enormous appeal for other leaders, especially less-impressive ones.

In international relations there is a theory that, especially in non-democracies, rulers will pursue international conflict, even war, as a way to divert attention from domestic political problems.[10] The Chávez case shows that the same factors can lead to a foreign policy of largesse, even the exportation of corruption.

### The First Paradox of Effective Social Power: Democracy Demotion

One would think that to project social power, all world leaders would need is money—in this case, petro-dollars. Venezuela is, after all, enjoying the most formidable oil boom since the 1970s; without it, the Chávez regime would be merely talk. But while money is crucial for the projection of social power, it is not the only requirement. Perhaps even more important than money is limited democracy at home. The reason is that social power requires freedom to spend unaccountably, and this is only possible when domestic institutions of checks and balances are disabled.

Social power as a foreign policy tool cannot flourish easily in a vibrant democracy. Foreign aid is always unpopular at home, where it competes with other spending priorities and domestic groups that have alternative plans for spending the money. Politicians running for office at home will want to devote more money to pork than to foreign spending. A system of governance that guarantees political competition and allows institutional space for opposition forces (which is what a democracy does) will naturally place checks on foreign profligacy.

Deploying a social-power foreign policy requires limiting the opportunities for the opposition to exercise influence, and even lessening the degree of political competition. The first paradox of a social-power foreign policy is that it is incompatible with democracy. Not surprisingly, therefore, the rise of social power in Venezuela has come with a decline in democracy.[11] Chávez essentially created what political scientists call a hybrid regime—one that is neither fully authoritarian, but not exactly democratic, either. Instead of abolishing checks-and-balances institutions, the regime packs them with loyalists. Instead of repressing dissidents, it practices job discrimination against voters. Instead of banning civic protests, it organizes counter-mobilizations by inciting and organizing mobs. Instead of disbanding organized opposition parties, it denies them resources. Instead of eliminating elected offices, it creates parallel, undemocratic units of government. Instead of shutting down the press, it burdens it with content regulations and, through media buyouts, reduces the private media's market share. Instead of suspending elections, it promotes abstentionism among would-be opponents by failing to guarantee the secrecy of the vote. Compared to the most repressive regimes of the twentieth century, the Chávez regime is relatively tame. But compared to most Latin American countries today, where indices of political and civil liberties are historically high, the Chávez regime is certainly the most intolerant of opposition, second only to Cuba.

### The Second Paradox: Who Cares about the Poor?

As a way to divert attention from its increasing authoritarian, conservative, militaristic, and business-collusion tendencies, the Venezuelan regime overemphasizes two features of its foreign policy: distributionism and anti-Americanism. Chávez wants to be recognized across the world as an undisputed champion of the poor. He spends far more on social programs than is typical of most authoritarian regimes, almost to the point where there are virtually no controls on spending. Chávez also wants to be known as a champion of anti-U.S. rhetoric. He estimates that as long as he can overplay these two features, progressive forces will either ignore or forgive the regime's less-than-progressive attributes. These features of *chavismo* will not go away. It is naive to believe that better U.S. diplomacy toward the regime will make it more fiscally frugal or less anti-American.

The Chávez government accomplishes its electoral victories by pursuing a classic populist strategy of inducements *and* constraints. Enormous benefits are allocated to those political actors willing to support the regime (e.g., access to new welfare programs, the so-called "missions," public-sector jobs, contracts with the state), while serious ostracism is applied to those who oppose it, mostly in the form of insults, job discrimination, legal accusations, and exclusion from welfare benefits. There is evidence that ostracism and concentration of power are on the rise, but the inducements continue as well. Given any petro-state's disproportionate advantage in controlling the mix of inducements and constraints to be dispensed, it becomes very difficult for any opposition force to beat the government in elections. The opposition has no means to reward supporters or punish opponents the way the state does.

Chávez's base of support is among low-income groups, but his coalition includes many high-income and privileged sectors of society. The most prominent non-disadvantaged ally is the military. After a series of purges between 2001 and 2004, the military is now well staffed with loyalists and larger than ever. Another government partner group is the so-called *boliburguesía*: business owners and business managers who support the Bolivarian revolution because they welcome state contracts and protectionism. Finally, Chávez is actively supported by those elements that profit from corruption and crime, especially the drug trade.

Chávez's social-power foreign policy mirrors his domestic policies. Hidden behind the sums of money allocated for the poor are real transfers to political elites who share Chávez's political objectives of expanding statism, imposing restrictions on the opposition, and undermining institutions of checks and balances.

## How the International Political Economy Fosters and Hinders the Exercise of Social Power

In addition to low levels of democracy, another factor aiding the rise of social power is a favorable international political economic environment.[12] Among the scholars who focus on the external causes of soft-balancing, the emphasis seems to be on the actions of the hegemon: the larger and more threatening the hegemon becomes, the more likely soft-balancing will emerge.[13] Favorable international conditions also seem to be necessary, however. High oil prices fuel Chávez's political excesses by providing the necessary revenues to expand statism, and thus the regime's policies of inducements and constraints.

A favorable IPE was a major contributing factor in Cuba's radicalization in the 1960s. The Soviet Union became willing to serve not just as Cuba's political sponsor, but also as its buyer of last resort, by absorbing sugar sales and providing subsidies (energy, capital, and financial assets). This allowed

Cuba to break trade ties (and thus political ties) with the United States and, one could even argue, disregard economic efficiency. However, it is important to keep in mind the differences in the international political economy of the early 1960s, when the Cuban Revolution became radicalized, and the late 2000s, when Venezuela's revolution could follow the same path. No country today is willing to play the Soviet Union's role. China, with its fast-growing appetite for oil, is the only likely candidate. Yet China is unlikely to offer to buy all of Venezuela's oil. In 2007, Venezuela sold approximately 1.23 million barrels per day of crude oil and petroleum products to the United States, and 0.35 million barrels per day of oil to China. Despite rhetoric from Caracas, this level of China-Venezuela oil trade is unlikely to grow significantly in the medium term. First, China knows that it is cheaper to develop markets near its border (Central Asia) or in countries where it can have greater bargaining leverage (African states). Furthermore, China does not have the refineries necessary to process Venezuela's extra-heavy crude. Shipping oil from Venezuela to China would also be prohibitively expensive, involving a forty-day trip.[14] Finally, China is unlikely to consider Venezuela to be the strong strategic asset that Cuba was for the Soviet Union. Back in the 1960s, acquiring a political/military base in Cuba was the equivalent of placing a battleship in the Gulf of Mexico and allowed the Soviet Union to balance NATO's military advantage in Europe (weapons near the Iron Curtain). China has no desire for a "battleship" in the Caribbean Basin.

Instead, Venezuela has turned to Iran. Not having found a buyer of last resort, Venezuela has developed a strong interest in maximizing the price of oil in world markets. Venezuela cannot accomplish this price hike alone, so it has become interested in Iran, one of the world's largest oil exporters. In addition to exchanging and developing weapons,[15] the primary purpose of ties with Iran is to push oil prices up.

There are two ways in which a Venezuela-Iran alliance could promote oil price increases. The first strategy is out in the open: increase the number of hawks within OPEC. Both countries are interested in price increases (through oil production cutbacks) rather than price stabilization (through production increases). Iran and Venezuela are the second- and fifth-largest producers in OPEC; partnership gives them the power within the oil cartel to counteract Saudi Arabia's attempts to keep production high. The other reason that an alliance with Iran could help promote price increases is a bit more sinister. Venezuela knows that a confrontation between Iran and the

United States could produce a crisis in the Middle East, which would boost the price of oil. If so, one could argue that Venezuela has an interest in encouraging such a crisis.

Table 2.1 shows the crucial actors in the world political economy of oil, as well as actual, tacit, and potential alliances. Venezuela and Iran share similar policy goals: to strengthen OPEC and oil prices. The United States and China share the opposite goals: to dilute OPEC's power and keep oil prices low. Saudi Arabia is the intermediate player. It sides with Venezuela and Iran on the issue of strengthening OPEC, but is somewhat sympathetic toward the United States and China on the issue of avoiding high oil prices. Insofar as the United States can keep China and Saudi Arabia on its side, it will preserve an international political economy that can contain some of Venezuela's foreign policy goals.

A confrontation between Iran and the United States could bring China closer to Venezuela. China has significantly increased its oil ties with Iran in the last decade. Iran is China's second-largest source of imported oil, and China dominates about 8 percent of Iran's oil market. If a confrontation between Iran and the United States were to occur, leading to an increase in the price of oil or a disruption in Iranian oil to China, the U.S.-Chinese alliance could become strained. At the very least, China would want to find new suppliers, and Venezuela could begin to look more appealing.

The reality, however, is that the United States depends far less on Venezuela than Venezuela does on the United States. Venezuela provides approximately 13 percent of U.S. oil imports, but the United States provides the bulk of Venezuela's export revenues (70 percent) and, therefore, its government revenues (almost 50 percent). Economically speaking, the United States is in

Table 2.1. Crucial Actors in the World Political Economy of Oil, and Actual, Tacit, and Potential Alliances

|  | Policy preference toward OPEC | Policy preference toward oil prices |
|---|---|---|
| Venezuela/Iran | strengthen | raise |
| Saudi Arabia | strengthen | moderate and stable |
| United States/China | weaken | lower |

a better position to absorb a hike in oil prices than Venezuela is to survive a collapse in oil sales to the U.S. Provided some conditions hold (inflation stays low, oil exporters continue to send their dollars to the United States), many analysts feel that the United States could survive further increases in oil prices.

Politically, for the United States unilaterally to end trade with Venezuela would be a serious public relations disaster and a boom for *chavismo*. Chávez is interested in a provocation with the United States. A confrontation would allow him to blame all his economic woes on the United States, concentrate more power, crack down on enemies by declaring a state of emergency, and gain even more international sympathy. Far from containing *chavismo*, a confrontation with the United States would embolden it.

### Chávez's Ineptness No. 1: Oil Crisis in Venezuela

Despite all the advantages that Chávez derives from using social power as a balancing foreign policy tool, he is far from a master at his own game. In particular, Chávez has consistently mishandled two factors that are necessary for exercising his brand of social power: he shows complete disregard for promoting efficiency in the oil sector, and he uses money abroad in ways that polarize rather than unite potential allies.

Chávez has failed to address or, one could argue, has even helped create a productivity crisis at PDVSA, the state-owned petroleum company. PDVSA is unquestionably the regime's vital cash cow. In late 2007, with international demand and prices for oil at their highest since 2003, one could have expected growing productivity within the Venezuelan petroleum sector. In fact, the opposite was true. The productivity of the petroleum sector dropped to record lows, mostly because of PDVSA's drop in productivity.[16]

Two principal factors explain this decline. The first stems from Chávez's decision, after the winter 2002–3 oil strike, to fire 20,000 PDVSA employees (almost 29 percent of the company's payroll), many of them technical experts or management personnel. The second factor is the replacement of multinational firms with transnational, state-owned oil firms. Chávez's petroleum foreign policy is split: on the one hand, he has opened the petroleum sector to *state-owned* petroleum enterprises from allied countries (China, Russia, Brazil); on the other hand, starting in 2005 he began to lift investment and operation barriers to the *private* international petroleum industry. The cost of this decision has been less foreign investment and domestic inefficiency.

Despite these efficiency holes, the political net balance is not necessarily negative. What Chávez loses in efficiency and investments he gains in terms of political control and discretionary power. The cracks in the system, however abundant, will not necessarily undermine the regime politically. The civilian opposition is numerically strong, but it lacks institutional mechanisms with which to act. So far, the opposition has rejected violence, which has saved it from repression but left it with few tools for pressuring the system. Ultimately, what keeps Chávez in office is his vast coalition of supporters. They find many faults with Chávez, but they also feel that with him as president their gains are guaranteed to last.[17]

### Bad Santa: Venezuela's Foreign Policy Blunders

Chávez has shown incompetence not just as a ruler, but also as a player at his own international game. Having discovered a potent foreign policy weapon—social power—he has been rather clumsy at exercising it. Chávez has done very little to hide the fact that he spends money on some foreign politicians and not on others. The overt political bias of this interventionism has two problems. First, it is not clear that Chávez has obtained support for some of his most-radical soft-balancing objectives. Other than ensuring a shield against criticism and a disregard for external conditionalities, his foreign policy influences its recipients' domestic politics more than their foreign policies.[18] Second, and more important, this foreign policy has brought angry responses from opposition politicians. Chávez has become enormously unpopular in the opposition parties of all the countries where he intervenes. Their denunciations have had much greater resonance than the criticisms of the U.S. government, international organizations, and even Venezuelan citizens. The best strategy to contain Chávez's social power, therefore, is to rely on existing democracies, because only democracies make room for strong opposition politics. The problem is that when social power is directed at a democratic regime, the incumbent becomes so empowered that even a scandalized and denunciatory opposition can do little to defeat the government politically.

In this context, Chávez can perhaps best be seen not as a "tropical Santa," as a former Venezuelan ambassador once put it, but as a "bad Santa": bad in the sense that he doesn't always deliver on what he promises, and that he polarizes almost every country in which he intervenes. In addition, his attempts to influence the politics of other countries sometimes backfire. In Peru in 2007 and Colombia in 2008, citizens ended up voting for the most anti-Chávez candidates on the ballot. Because he wants to use social spending to create clones abroad, not just a diplomatic shield, Chávez suffers numerous diplomatic setbacks.

The advantage of the role of international Santa, however, even if the role is played badly and abusively, is that the world community seldom reacts with outright condemnation. Santa may disappoint, but he never appears too threatening to others, except to opposition politicians in the countries where he intervenes. In terms of public relations, the misuse of social power is therefore less damaging than the misuse of hard power, which tends to produce more virulent international condemnation. Herein lies the advantage of Venezuela's foreign policy tool.

### The Trade-Offs of Chávez's Foreign Policy

It would be disingenuous to think that social power is Chávez's only foreign policy objective. Like most nations, Venezuela pursues multiple foreign policy goals. The problem is that many of these goals contradict and counteract each other.

Table 2.2 summarizes some of Chávez's most important foreign policy objectives, in addition to social power. Pursuing any one of these policies (indicated with a checkmark) will frustrate one of the other goals (marked with an "X").

For instance, Venezuela would like to reduce its dependence on the U.S. as an export market and at the same time disrupt the U.S. oil supply. However, this would entail sacrificing distributionism and social power as a foreign policy goal, since Chávez has no alternative market for its oil or, therefore, alternative source of revenue. Venezuela has also expressed interest in developing the Chinese oil market. This would require heavy investments in refineries and shipping costs, which would undermine distributionism as well as Venezuela's desire to keep oil prices high (since developing the Chinese market would ease pressure on supply restraints). Venezuela wants to pursue

Table 2.2. Venezuela's Foreign Policy Goals and Trade-Offs

| | End dependency on U.S. oil market | Develop Chinese market for Venezuelan oil | Expand social spending/ Consumer boom | Raise oil prices/ Alliance with Iran | Strengthen OPEC | Expand social power diplomacy | Promote state-owned enterprises | Promote investments at home/ Raise productivity | Intensify soft balancing/ Conflict with Colombia | Intervene in elections abroad | Unite the region | War with Colombia |
|---|---|---|---|---|---|---|---|---|---|---|---|---|
| End dependency on U.S. oil market | ✓ | | X | | | X | | | | | | |
| Develop Chinese market for Venezuelan oil | | ✓ | | X | | X | | X | | | | |
| Expand social spending/Consumer boom | | | ✓ | | | | | X | | | | X |
| Raise oil prices/Alliance with Iran | | | | ✓ | X | | | | | | X | |
| Strengthen OPEC | | | | | ✓ | | | | | | | |
| Expand social power diplomacy | | | | | | ✓ | | X | | | X | |
| Promote foreign investments by state-owned enterprises (rather than private firms) | | | | | | | ✓ | X | | | | |
| Raise productivity of oil sector | | | | | | | | ✓ | | | | |
| Excessive soft-balancing of the U.S./Support radicals/Cuba | | | | | | | | | ✓ | | X | |
| Intervene in elections abroad | | | | | | | | | | ✓ | X | |
| Unite the region | | | | | | | | | | | ✓ | X |
| War with Colombia | | | | | | | | | | | | ✓ |

Note: ✓'s and X's illustrate how pursuit of any particular goal is balanced by the frustration of other goals.

high oil prices, but this would cause rifts with China and OPEC. Venezuela may be tempted to seek international confrontations for the sake of rallying domestic support, but this would undermine Chávez's capacity to project social power. Chávez has a penchant for dealing with state-owned, rather than private-owned, oil multinationals, but this impairs his ability to acquire the capital and technology needed to keep the oil sector productive and, in turn, for him to pursue his foreign policy objectives.

Two main conclusions can be drawn from Table 2.2. First, we should expect erratic foreign policies from Venezuela. One day Chávez may prefer one of these objectives, but he is likely to desist soon, given the tradeoffs involved. Second, of all the foreign policy objectives he may pursue, projecting social power is likely to persist because it seems to entail the fewest tradeoffs and costs.

## How to Meet the Social Power Challenge

Hugo Chávez has discovered an innovative foreign policy instrument: heavy international spending with a veneer of subsidizing the poor. This type of power, which is different from traditional hard power or soft power, allows the regime to shield itself from diplomatic criticism and gain admirers. It also influences elections abroad. The Venezuelan case proves that social power as a foreign policy tool to attract goodwill can work, even when exercised incompetently. Furthermore, precisely because it is such a friends-making, low-risk foreign policy tool, future U.S. rivals can easily emulate it and, with a little effort, become better players at it than Chávez. It is a form of diversionary foreign policy that appeals to the most unappealing regimes.

The United States must develop a counterstrategy to Venezuela's "social power" diplomacy. A "hard power" response—military or economic aggression—seems disproportionate to the offense. A "soft power" response—preaching the virtues of liberal democracy, for example—would have little impact on illiberal political movements and would not lessen the demand for foreign aid that Venezuela satisfies.

Emulating Venezuela—that is, offering more foreign aid—is not the solution either. The United States already devotes a large amount of aid to Latin America. Adding substantially to this pool may have little marginal return. Part of the problem with U.S. aid is that, while substantial, it is generally indirect (offered through various agencies) or comes with strings attached. The United States cannot simply become the charity organization of last resort. Adding more aid would not diminish the demand for the type of foreign policy that Venezuela projects, and so would not deal with the problem of social power by balancing regimes. Making aid less conditional could also undermine the U.S. goals of democracy and governance promotion.

The challenge for the United States in dealing with Venezuela is not necessarily (or at least not yet) how to contain the Venezuelan threat, but how to respond to a foreign policy tool that is so suitable for an age in which petro-states have become major world actors, incentives to soft-balance the United States are strong, and advocates around the world are calling for unconditional development aid.

## Notes

1. Joseph S. Nye, "Soft Power and American Foreign Policy," *Political Science Quarterly* 119, no. 2 (2005): 255–70.

2. Robert A. Pape, "Soft Balancing against the United States," *International Security* 30, no. 1 (2005): 7–45; T. V. Paul, "Soft Balancing in the Age of US Primacy," *International Security* 30, no. 1 (2005): 46–71; Andrew Hurrell, "Hegemony, Liberalism, and Global Order: What Space for Would-be Great Powers?" *International Affairs* 82, no. 1 (2006): 1–19.

3. Mark E. Williams, "International Relations Theory and Venezuela's 'Soft-Balancing' Foreign Policy," in *The Revolution in Venezuela*, ed. Jonathan Eastwood and Thomas Ponniah (Chapel Hill, N.C.: Duke University Press, 2007); Mark E. Williams, "The New Balancing Act: International Relations Theory and Venezuela's Soft-Balancing Foreign Policy," *Latin American Studies Association* (forthcoming).

4. Richard Feinberg, "Chávez Conditionality," Perspectives on the Americas Series (Center for Hemispheric Policy, University of Miami, 2007).

5. Sean W. Burges, "Building a Global Southern Coalition: The Competing Approaches of Brazil's Lula and Venezuela's Chávez," *Third World Quarterly* 28, no. 7 (2007): 1343–58.

6. See Maingot, "Responses to Venezuelan Petro-politics in the Greater Caribbean," this volume.

7. Janet Kelly and Carlos A. Romero, *The United States and Venezuela* (New York: Routledge, 2002); María Teresa Romero, *Política exterior venezolana* (Caracas: Editorial El Nacional, 2002); Demetrio Boersner, "Dimensión internacional de la crisis venezolana," in *Venezuela en retrospectiva: Los pasos hacia el regimen chavista*, ed. Günther Maihold (Madrid: Iberoamericana; Frankfurt: Vervuert, 2007).

8. Jorge I. Domínguez, *To Make a World Safe for Revolution: Cuba's Foreign Policy* (Cambridge, Mass.: Harvard University Press, 1989).

9. See Buxton, "European Progressives and the Bolivarian Social Agenda," this volume.

10. Jack Levy and Lily I. Vakili, "Diversionary Action by Authoritarian Regimes: Argentina in the Falklands/Malvinas Case," in *The Internationalization of Communal Strife*, ed. Manus I. Midlarsky (New York: Routledge, 1992), 118–46; Graeme A. M. Davies, "Domestic Strife and the Initiation of International Conflicts: A Directed Dyad Analysis, 1950–1982," *Journal of Conflict Resolution* 46, no. 5 (2002): 672–92.

11. In addition to a decline in democracy, there has been a decline in the professionalism of the diplomatic corps. In April 1999, Chávez ordered a "gigantic shuffling of diplomatic and consular personnel abroad," transferring many people back home and removing key staff. Nearly fifty diplomatic missions lost their leaders (see Demetrio Boerner, "Dimensión internacional de la crisis venezolana," in *Venezuela en retrospectiva: Los pasos hacia el régimen chavista*, ed. Günther Maihold (Madrid: Iberoamericana; Frankfurt: Vervuent, 2007, 316–17).

12. This section and the next draw from Javier Corrales, "The Venezuelan Political Regime Today: Strengths and Weaknesses," in *Proceedings of the 8th Conference on U.S. Policy in Latin America* (Washington, D.C.: The Aspen Institute, 2007), 13–19.

13. Others would suggest that soft-balancing, or balancing in general, is a response to domestic factors. According to the theory of "diversionary" conflict, balancing in general, even provoking war, can be caused by domestic political insecurities under certain fiscal pressures and military conditions.

14. Because supertankers are not allowed through the Panama Canal, oil shipments from Venezuela to China would need to go first south-southeast to the Strait of Magellan and then north-northwest across the Pacific, or entirely east through Cape Horn and then the Strait of Malacca. Either route would be one of the longest in the world.

15. There is some concern that Iran is helping Venezuela explore the nuclear option. There are rumors that Iranian scientists and engineers are prospecting for uranium ore in the granite bedrock under the jungles of southeastern Venezuela, a region rich with mineral deposits. It is difficult to see why Chávez would want nuclear technology for peaceful, energy-producing ends: Venezuela has the largest hydrocarbon reserves in the Americas, and it already makes good use of its ample hydroelectricity generation potential.

16. In 2003, at the time of the petroleum stoppage, PDVSA's productivity reached 2.2 million barrels per day (not including the production of hydrocarbons by foreign businesses). In late 2007, PDVSA was producing about 1.6 million barrels a day—30 percent less than in 2003, although it is difficult to say with certainty, since the company will not provide definite and verifiable data. In September 2007, OPEC reassigned its members' quotas to better reflect their current production. Venezuela's quota was reduced from 11.5 percent to 9.1 percent (from 3,085,000 barrels per day, to 2,470,000 barrels per day) as of November 1, 2007, information that the Chávez government tried to hide but which came to public light.

17. In December 2007 a government-sponsored referendum for constitutional change

aimed at reducing checks and balances on the president was defeated by a narrow margin of 1.4 points. This was Chávez's first defeat at the polls. One reason was a high abstention rate among *chavistas*, which suggests that not all *chavistas* favor uncontrolled expansion of the president's formal powers.

18. Sean W. Burges, "Building a Global Southern Coalition: The Competing Approaches of Brazil's Lula and Venezuela's Chávez," *Third World Quarterly* 28, no. 7 (2007): 1343–58.

# 3

# Public Opinion and Venezuelan Foreign Policy

JOHN MAGDALENO G.

Many observers have noted the assertive nature of Venezuela's foreign policy, the efforts by the Venezuelan government to market its image internationally, and the impact of this strategy on target audiences. An equally important consideration, however, is the progressive, albeit gradual, shift in opinion in Venezuela of Hugo Chávez's foreign policy. How do Venezuelans view the evolution of this policy? How has it affected overall public opinion of the president? Finally, how can we evaluate a foreign policy that has had increasing costs abroad and well as domestically?

To answer these questions, in the following discussion I examine the empirical evidence provided by a sample of public opinion surveys with national or urban national coverage dating back to 2002. The discussion is highly descriptive, a common characteristic of exploratory analyses of quantitative surveys, but it also suggests hypotheses and possible interpretations of the results presented, as well as a number of observations on future prospects for Chávez's foreign policy.

## Perceptions of Venezuelan Foreign Policy

A February 2002 survey by DatAnalysis, a leading Venezuelan polling firm, compared Venezuelans' perceptions of the course of Chávez's foreign policy with the direction they would like to see this policy take (see Figure 3.1).[1] The results are clear; while 53 percent of those surveyed said that the Chávez government should "foment relations with all the countries in the world" without regard for ideology, 63 percent perceived that "relations with non-

## What you think is

| Category | Percentage |
|---|---|
| Foment relations with non-democratic countries or state-controlled economies like Cuba, Libya, Iraq, Iran, and Saudi Arabia | 63.3% |
| Foment relations with all the countries in the world, without any of the other two views prevailing | 14.5% |
| Foment relations with democratic and market economies like the United States, countries in Europe, and some countries in Latin America | 13.1% |
| Don't know | 7.1% |
| No Answer | 2.0% |

## What you think should be

| Category | Percentage |
|---|---|
| Foment relations with non-democratic countries or state-controlled economies like Cuba, Libya, Iraq, Iran, and Saudi Arabia | 3.2% |
| Foment relations with all the countries in the world, without any of the other two views prevailing | 52.7% |
| Foment relations with democratic and market economies like the United States, countries in Europe, and some countries in Latin America | 39.7% |
| Don't know | 3.2% |
| No Answer | 1.2% |

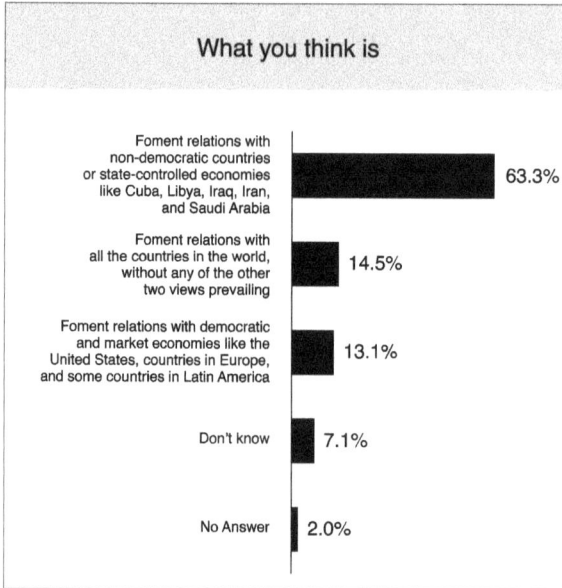

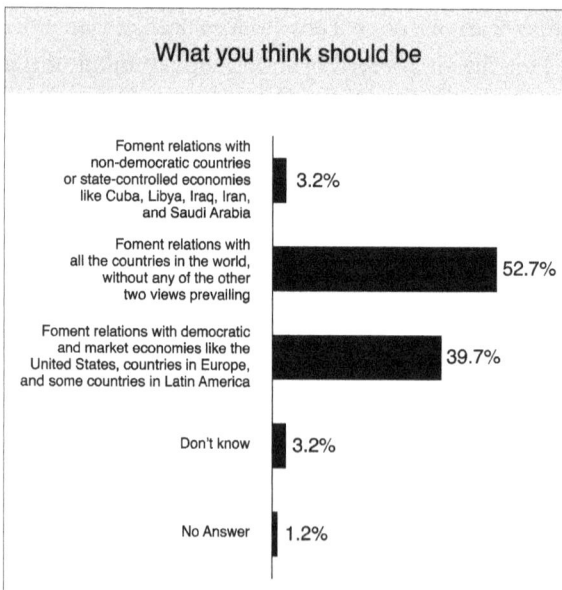

Figure 3.1. Perceptions of the Direction of President Chávez's Foreign Policy

1. Study conducted among men and women 18 and older in socioeconomic levels A/B, C, D, and E in Venezuela's 25 main urban areas. The sample size was 1,000, with a maximum allowable error of +3.04 at the 95% confidence level. The sample was collected using non-probability dynamic quotas (controlling for gender, age, and socioeconomic level). Interviews were conducted at home, face-to-face, on February 22–26, 2002.

democratic countries or state-controlled economies such as Cuba, Libya, Iraq, Iran and Saudi Arabia" were in fact the ones being fomented.

It is worth noticing that 40 percent of those surveyed answered that they preferred a foreign policy concentrated on strengthening "relations with democratic and market economy countries such as the United States, countries in Europe and some countries in Latin America," but only 13 percent perceived that this was the principal foreign policy direction at the time. We can say, therefore, that as of February 2002 Venezuelan foreign policy was heading in a direction away from the preferences of the majority of those interviewed.

A similar study in September 2005, also by DatAnalysis, included an additional question that attempted to measure which scenario interviewees considered more probable (see Figure 3.2).[2] A majority (56.5 percent) of those interviewed preferred a future foreign policy directed at strengthening relations "with all the countries in the world without a particular view prevailing," but only 34 percent considered that scenario probable. Twenty-eight

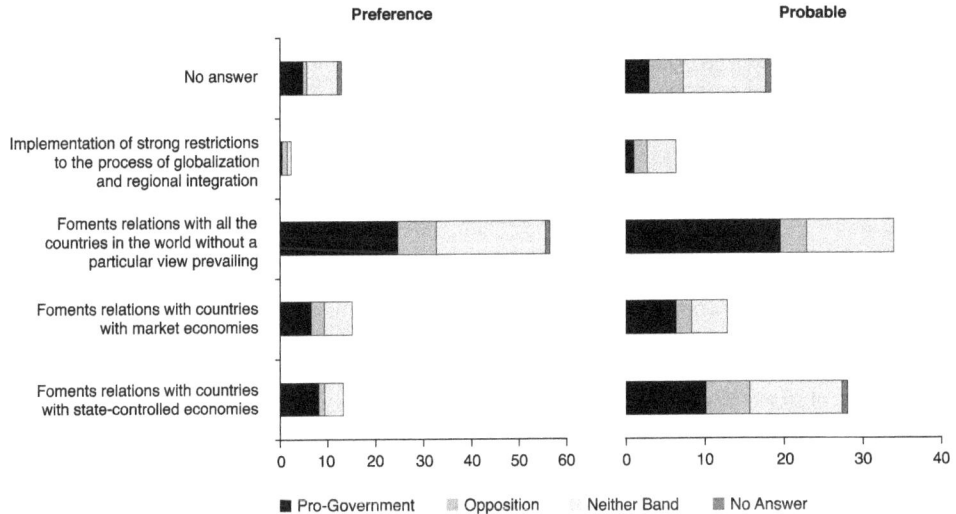

Figure 3.2. *Which international relations model would you prefer for the year 2015, and which do you think is most probable?*

1. Study conducted among men and women 18 and older in socioeconomic levels A/B, C, D, and E in 50 localities across the country. The sample size was 1,300, with a maximum allowable error of +2.7% at the 95% confidence level. The sample was collected using non-probability dynamic quotas (controlling for gender, age, and socioeconomic level). Interviews were conducted at home, face-to-face, on September 16–25, 2005.

percent believed that it was probable that in the year 2015 Venezuelan foreign policy would foment "relations with countries with state-controlled economies," although only 13 percent indicated this as their preference. Another 13 percent believed it was probable that the Venezuelan government would foment "relations with countries with market economies," versus 15 percent who mentioned this as their preference (the smallest gap observed to date).

If we contrast the September 2005 results with perceptions in February 2002 about the real direction of Venezuelan foreign policy, some variations become evident (see Figure 3.3). Thirty-four percent of respondents said that the current government "foments relations with countries with state-controlled economies"; 28 percent said that the government "foments relations with all the countries in the world without a particular view prevailing"; 15 percent indicated that the government "foments relation with countries with market economies"; and 7 percent selected a new option: "Implementation of strong restrictions on the process of globalization and regional integration." While the wording in this survey is not identical to that used in the February

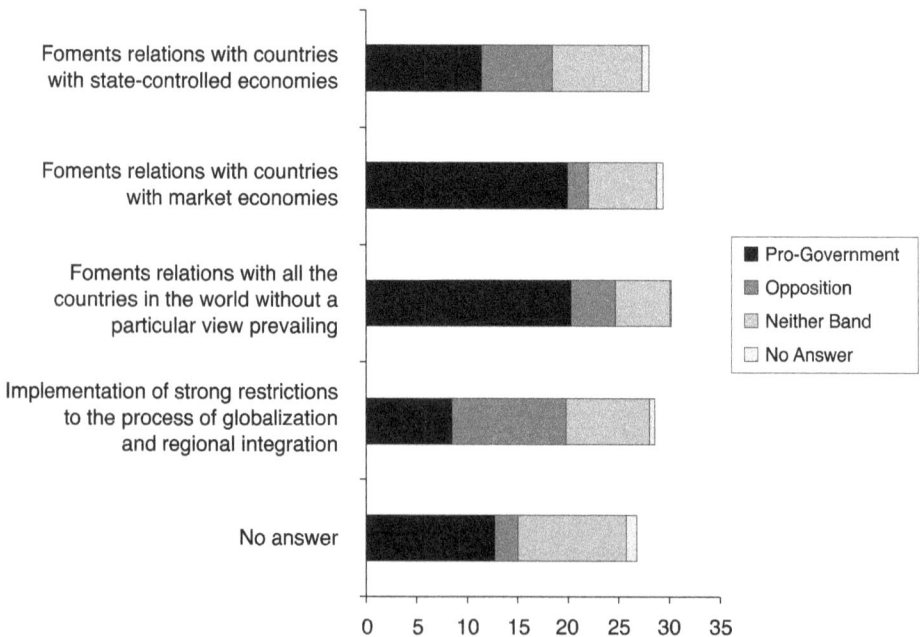

Figure 3.3. *What do you think is the government's current international relations model?* (Values expressed as percentages)

2002 poll, there is a relevant variation of nearly 30 percentage points for the first type of answer.

Two hypotheses can be formulated from this analysis. First, the September 2005 poll came one year after the presidential revocation referendum of August 2004—which was followed in October by regional and local elections—and just two months before parliamentary elections. Most opposition parties were urging people not to vote, to protest what they considered fraud in the revocation referendum. In a study conducted by DatAnalysis shortly after the referendum, 39 percent of those surveyed stated that the consultation had been fraudulent, and 4 out of 10 respondents expressed political apathy. If the types of answers in September 2005 were identical to those in the 2002 study and Figure 3.3 reflects a typing or transcription error, then it may be possible that the postelectoral climate that followed the announcement of the referendum results influenced the interviewees' perceptions of Venezuelan foreign policy. A second hypothesis is that the results were closely related to the increasing influence of the official media, especially after the restrictions that began in 2004 on private media who followed an editorial line against the government. Under this hypothesis, the 30 percent gap between February 2002 and September 2005 with regard to Venezuela's foreign relations with non-democratic countries with state-controlled economies may have been related to the government's success in marketing a different vision of its foreign policy to the majority of the population.

The two hypotheses are not mutually exclusive, and may both explain the variation recorded. If "Venezuela's assertive global projection" is to be understood as the result of the efficacy of government communication, in Venezuela and abroad, then this could be true only if two essential variables are taken into account: first, the increase in Venezuelan oil income from higher world oil prices, which allowed the government to finance information campaigns abroad, court government-friendly "opinion makers," and hire lobbying firms in countries such as the United States; and second, the progressive construction of a "communications and propaganda machine" at the service of the Venezuelan state to legitimize the official version of events, market the Chávez government, and filter or block access to official media by political or social actors critical of the government.

The survey results are tentative evidence that, in the last quarter of 2005, the government was able to exert a greater influence on Venezuelan public opinion, particularly on Venezuelans' perceptions of foreign policy. An important gap remained between desired and perceived foreign policy, but the discrepancy had lessened since February 2002.

### Perceptions of Cuba and the United States

The Chávez government has a history of using the United States, and particularly former president George W. Bush, as a straw man, but do most Venezuelans share this negative view? Chávez has made no secret of his anti-Americanism,[3] or of his relationship with Fidel Castro, whom he has publicly vindicated in speeches, official visits, trade agreements, and by means of an oil supply agreement at preferential prices. It is worth knowing how Venezuelans value these relations. The time series in Figure 3.4 clearly shows consistently high levels of rejection to this question.[4] Nevertheless, support

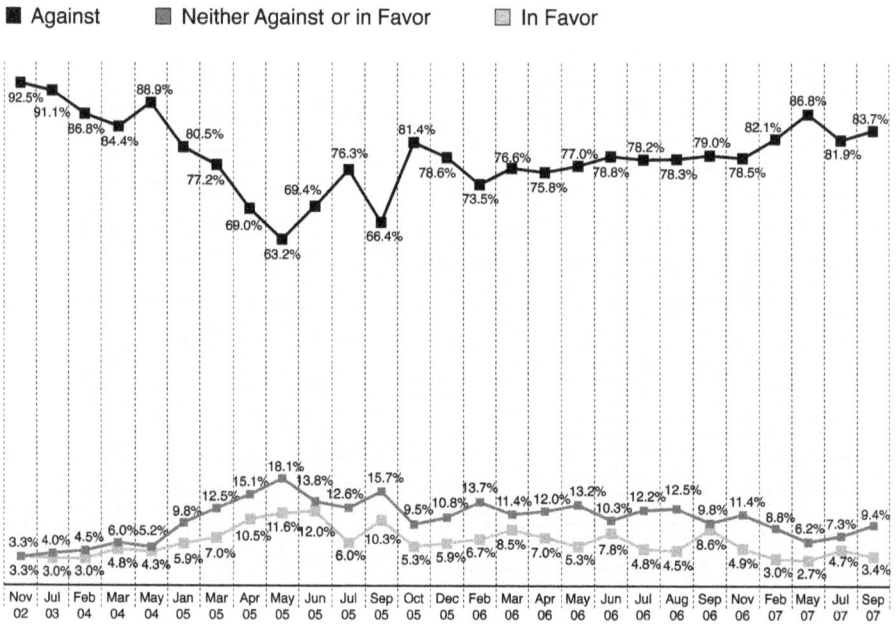

Figure 3.4. *What is your position with regard to Venezuela taking Fidel Castro's Cuban regime as an example?*

1. The question was part of DatAnalisis's Omnibus Survey, with the same sample design as the one described in Figure 3.2. The Omnibus Survey had a sample size of 1,000 up to July 2003, and 1,300 thereafter, with the exceptions noted in Figure 3.4.

for the Cuban model varied significantly over the sampling period. Negative responses dropped significantly, from 93 percent to 63 percent, between November 2002 and May 2005, and increased between July 2005 and September 2007. One explanation could be the successful efforts by the Chávez government to influence Venezuelan public opinion during these two periods.

In late 2004, after victories in the revocation referendum and regional elections, Chávez embarked on a strategy of political persuasion that emphasized gradual legitimization of the Cuban model. The strategy was simple: to promote the benefits of Misión Barrio Adentro (Mission Inside the Neighborhood), the government's flagship social program for the poorest segments of the population. Misión Barrio Adentro is a primary health care network staffed by Cuban doctors at small, government-provided clinics in poor areas of the country. On national television and radio, Chávez praised Cuba's contributions to the "Bolivarian revolution" and the poorest Venezuelans. For example, on the June 12, 2005, broadcast of *Aló, Presidente*, his regular Sunday television show, Chávez focused on three main themes: advances in the Bolivarian revolution's health programs, the brotherhood between Cuba and Venezuela, and the legitimization of socialism based on reaffirmation of the capitalist-socialist dichotomy, expressed as the confrontation between neoliberalism and "twenty-first century socialism." These three topics can be synthesized into one—the legitimization of socialism through health care.

Content analysis of the words Chávez used in that program, and especially a frequency count of key words, shows the hammering home of the healthcare-Cuba-socialism chain. Chávez used the words "Cuba," "Cuban" or "Cubans" 185 times, the third-highest frequency of any of the words he used that day (after "Venezuela" or "Venezuelans"—244 times—and "I"— 224 times). Other high-frequency words included "people" and "peoples" (123 times), "revolution" or "revolutionary" (54 times), "Bolívar" (45 times), "socialism" (43 times), "capitalism" (13 times), "poor" or "poverty" (19 times), and "middle class" (14 times).[5] Chávez repeated positive messages regarding Cuba and Fidel Castro on many programs from 2003 through 2005.

The question follows, therefore, why the surveys found increased rejection of the Cuban model after July 2005, especially in light of the president's public relations effort and a favorable public opinion climate after three successful elections between 2004 and 2005. A tentative explanation is that this rejection was associated with the progressive introduction of radical ideo-

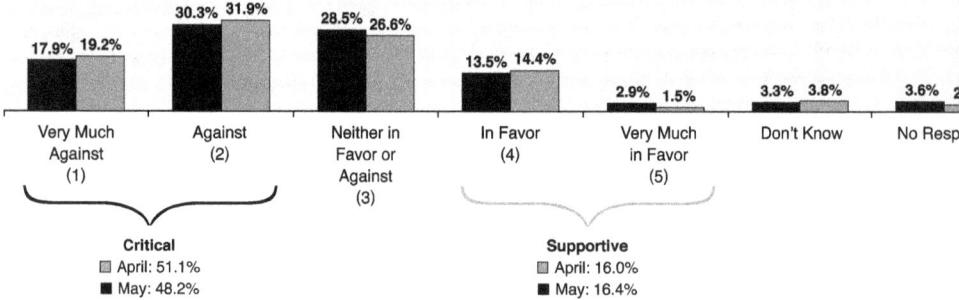

Figure 3.5. *What is your position with regard to the government of Cuba, presided over by Fidel Castro?*
    1. This study followed the same sample design as Figure 3.2.

logical content into Chávez's discourse. Starting around this time, the president made frequent references to Marx, Engels, Lenin, and Trotsky in his public comments, and insisted on collective modes of labor organization. Many Venezuelans may have interpreted these references as a sign of radicalization, and felt unhappy at the prospect.

DatAnalysis conducted two additional public opinion surveys with national urban coverage in April and May 2005, in which subjects were asked their opinion of Fidel Castro's government (see Figure 3.5).[6] Even when rejection of the Cuban model as an example for Venezuela was at its lowest levels (although it still accounted for the majority of the population), rejection of Fidel Castro's government exceeded 50 percent. Even among Chávez supporters, only slightly more than one-third had a favorable opinion of the Cuban government.

In the same survey, interviewees were asked whether the United States should be used as an example for Venezuela (see Figure 3.6). Clearly, overwhelming rejection of the Castro regime did not translate automatically into approval of the United States. In April 2005, rejection of the U.S. model was 40 percent, rising to 45 percent in May. Using the latter date as a reference, the most significant difference between opinion segments is that the United

Base: 1,300

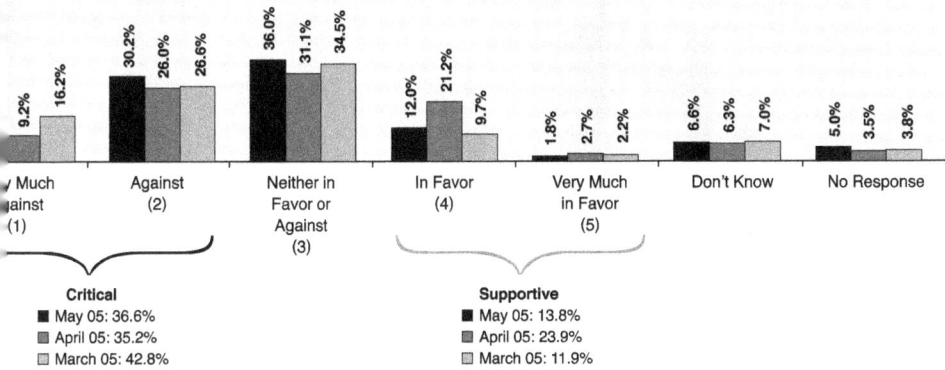

| | Critical | | | Supportive | |
| --- | --- | --- | --- | --- | --- |
| | May 05: 36.6% | | | May 05: 13.8% | |
| | April 05: 35.2% | | | April 05: 23.9% | |
| | March 05: 42.8% | | | March 05: 11.9% | |

Figure 3.6. *What is your position with regard to Venezuela taking the United States government as an example?*
1. This study followed the same sample design as Figure 3.2.

States seemed like an attractive model for about half of interviewees who defined themselves as opposing the Chávez government.

Between March and May 2005, DatAnalysis surveys also asked Venezuelans their opinion of the government of President George W. Bush (see Figure 3.7). Between April and May 2005, rejection of the Bush government oscillated between 35 percent and 43 percent, depending on the measurement at hand, with support ranging between 12 percent and 24 percent. Responses that were "neither in favor nor against" remained more consistent, between 31 percent and 36 percent. Therefore, just as rejection of the Cuban model did not necessarily mean support for a U.S. model for Venezuela, rejection of Fidel Castro's government did not automatically translate into approval for George W. Bush's government. Based on the empirical evidence presented, we can conclude that neither government was attractive to Venezuelans, even though more rejected the Castro government than did the U.S. government.

A clue to these opinions about Cuba and the United States can be found in Figure 3.8. Thirty-nine percent of interviewees preferred the U.S. model as a guide for Venezuela in May 2005 (12 percentage points less than in April 2005), compared to 15 percent who preferred the Cuban model (a difference

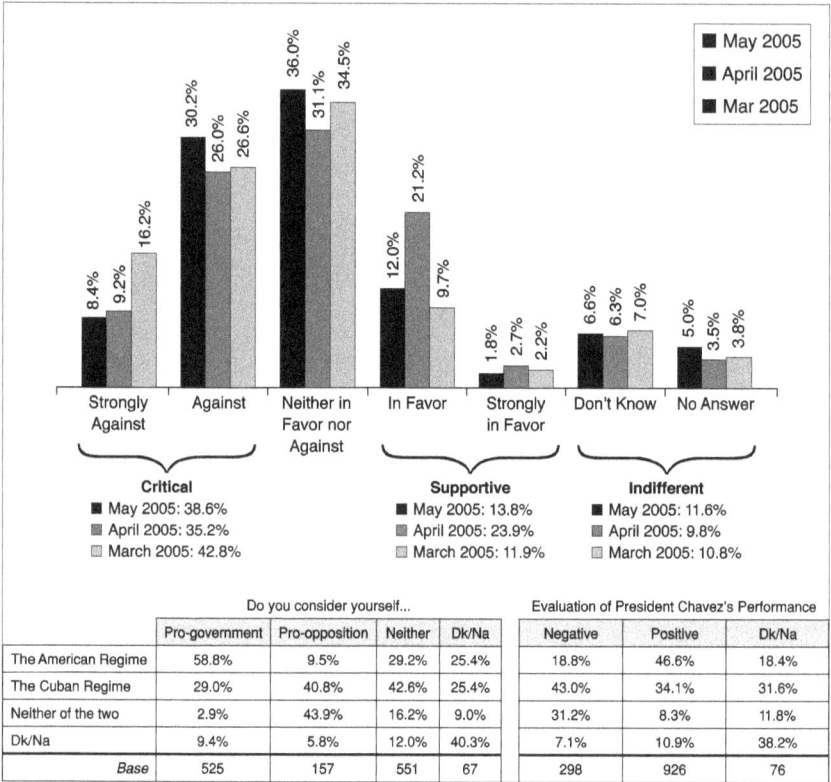

Figure 3.7. *What is your position with regard to the government of the United States, presided over by George W. Bush?*

1. Base: 1,300.

of only one percentage point, signaling a more stable opinion). Twenty-six percent answered "neither of the two"; when added to those who answered "don't know" or "no answer," this means that 59 percent of interviewees preferred neither model.

It is worth emphasizing that the loss of support for the U.S. model between April and May 2005 did not translate into more support for the Cuban model. It did mean, however, an increase in the opinion that neither model was a guide for Venezuela. This finding is highly relevant in the current sociopolitical context, suggesting a number of interpretations.

Despite the political polarization dating back to 1999—a polarization that carries over to opinions of Cuba and the United States, given the constant

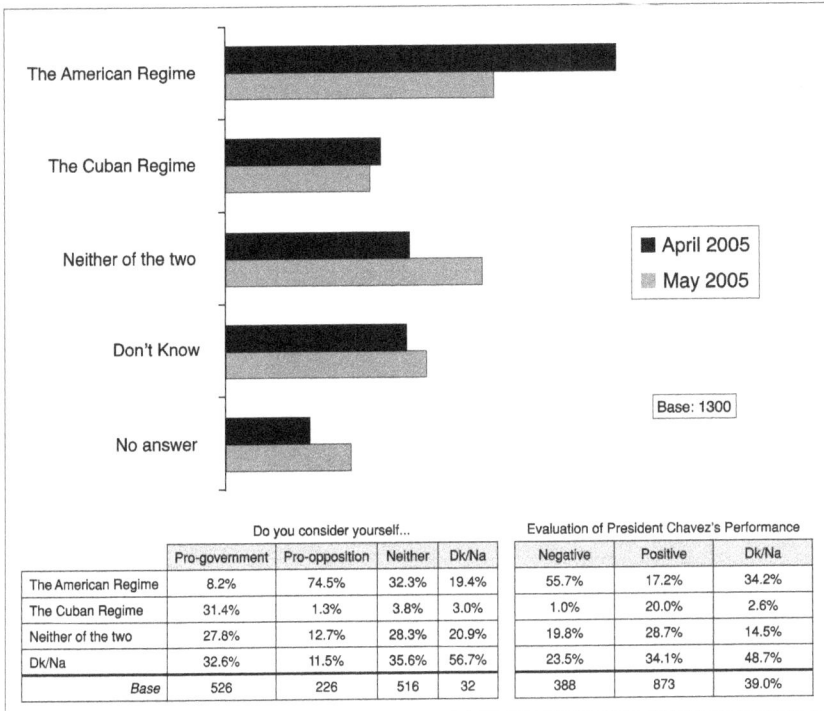

| Do you consider yourself... | | | | | Evaluation of President Chavez's Performance | | |
| | Pro-government | Pro-opposition | Neither | Dk/Na | Negative | Positive | Dk/Na |
| --- | --- | --- | --- | --- | --- | --- | --- |
| The American Regime | 8.2% | 74.5% | 32.3% | 19.4% | 55.7% | 17.2% | 34.2% |
| The Cuban Regime | 31.4% | 1.3% | 3.8% | 3.0% | 1.0% | 20.0% | 2.6% |
| Neither of the two | 27.8% | 12.7% | 28.3% | 20.9% | 19.8% | 28.7% | 14.5% |
| Dk/Na | 32.6% | 11.5% | 35.6% | 56.7% | 23.5% | 34.1% | 48.7% |
| Base | 526 | 226 | 516 | 32 | 388 | 873 | 39.0% |

Figure 3.8. *If you had to choose between the Cuban regime and the American regime as a guide for Venezuela, which one would you choose?*

references by political actors to these opposing models—the majority of Venezuelans were looking for a "third option." This suggests a level of "wear and tear" for these two models among the majority of interviewees, especially the U.S. model, even though it remained comparatively more attractive than the Cuban model. This decline can be associated with Chávez's influence over public opinion, and with the stereotypes, prejudices, and social representations cited so frequently in his speeches. Finally, lower levels of support for the U.S. model indicates a failure by U.S. political and economic actors to legitimize themselves and their model among Venezuelans.

Loss of support for the U.S. model did not entail approval for an eventual confrontation with the United States, however, as threatened so often in Chávez's speeches (see Figure 3.9). Approximately 80 percent of Venezuelans disagreed with Chávez's warnings of an eventual armed confrontation with the United States. Most striking is the high level of rejection of this proposal

**President Chávez says we are prepared for a war with the United States**

### September 2006

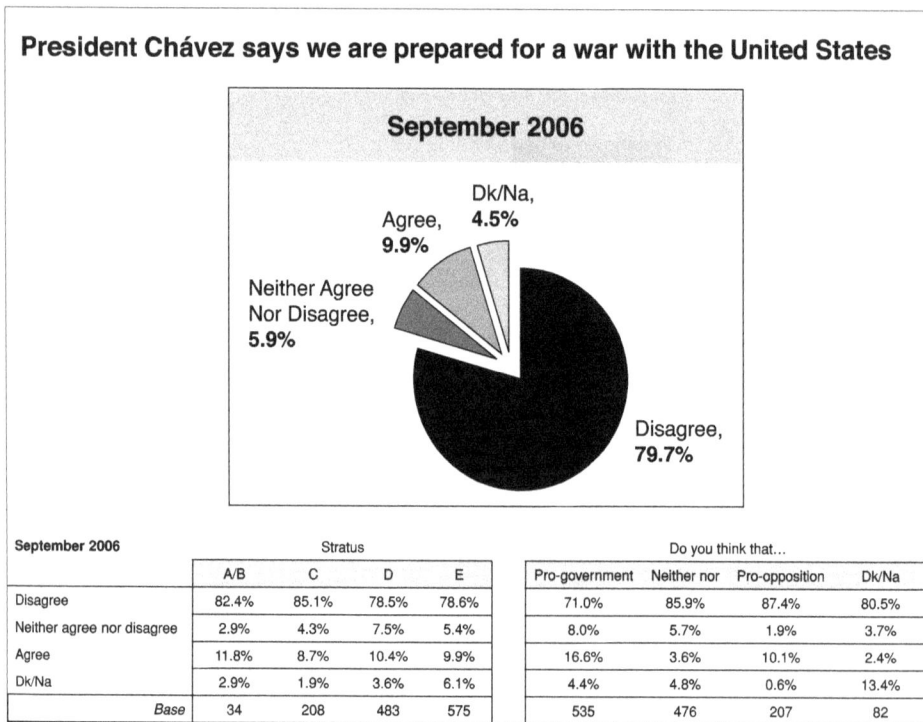

Dk/Na, **4.5%**

Agree, **9.9%**

Neither Agree Nor Disagree, **5.9%**

Disagree, **79.7%**

| September 2006 | Stratus | | | | Do you think that... | | | |
|---|---|---|---|---|---|---|---|---|
| | A/B | C | D | E | Pro-government | Neither nor | Pro-opposition | Dk/Na |
| Disagree | 82.4% | 85.1% | 78.5% | 78.6% | 71.0% | 85.9% | 87.4% | 80.5% |
| Neither agree nor disagree | 2.9% | 4.3% | 7.5% | 5.4% | 8.0% | 5.7% | 1.9% | 3.7% |
| Agree | 11.8% | 8.7% | 10.4% | 9.9% | 16.6% | 3.6% | 10.1% | 2.4% |
| Dk/Na | 2.9% | 1.9% | 3.6% | 6.1% | 4.4% | 4.8% | 0.6% | 13.4% |
| Base | 34 | 208 | 483 | 575 | 535 | 476 | 207 | 82 |

Figure 3.9. *President Chávez says we are prepared for a war with the United States. How strongly do you agree with this statement?*
   1. This study followed the same sample design as Figure 3.2.

among government supporters (71 percent), in addition to the percentages recorded for those who identified themselves as "non-aligned" and "opposition" (86 percent and 87 percent, respectively).[7]

## Foreign Spending and the FARC

Another important element of Chávez's foreign policy is his largesse when courting sympathetic governments. A September 2006 DatAnalysis survey asked Venezuelans how they felt about this policy (see Figure 3.10). About 70 percent of the respondents disagreed with their government's spending in other countries. This opinion was dominant among all socioeconomic strata and the three main opinion segments, including those friendly to Chávez and the Venezuelan government.

## President Chávez gives money away to other countries

### September 2006

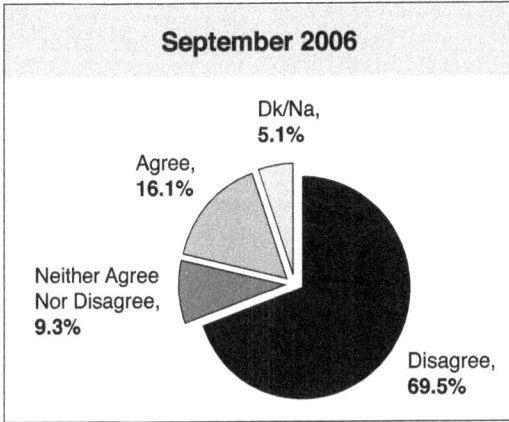

Dk/Na, 5.1%

Agree, 16.1%

Neither Agree Nor Disagree, 9.3%

Disagree, 69.5%

| September 2006 | Stratus | | | | Do you consider yourself... | | | |
|---|---|---|---|---|---|---|---|---|
| | A/B | C | D | E | Pro-government | Neither | Pro-opposition | Dk/Na |
| Disagree | 79.4% | 75.5% | 67.1% | 68.9% | 62.1% | 75.6% | 78.3% | 61.0% |
| Neither agree nor disagree | 8.8% | 8.7% | 11.0% | 8.2% | 12.3% | 9.0% | 7.4% | 85.0% |
| Agree | 8.8% | 13.0% | 16.1% | 17.6% | 20.9% | 9.2% | 18.8% | 17.1% |
| Dk/Na | 3.0% | 2.8% | 5.8% | 5.3% | 4.7% | 6.2% | 0.5% | 13.4% |
| Base | 34 | 208 | 483 | 575 | 535 | 476 | 207 | 82 |

Figure 3.10. *How strongly do you agree with . . . ?*

Another DatAnalysis study, between July and August 2007, asked intervie-wees about their level of agreement or disagreement with specific measures taken by the Venezuelan government to favor other countries (see Figure 3.11).[8] The two phrases related to financing infrastructure or making dona-tions to other countries accounted for more than 60 percent of disagreement among interviewees (61 percent and 64 percent, respectively), in addition to statements about the purchase of weapons and warnings of an eventual armed conflict with the United States (63 percent and 75 percent). The high-est level of rejection was for the use of public funds to support "the recon-struction of homes and military bases in some municipalities in Bolivia," except among respondents friendly to the government.[9]

Another study, between February and March 2006, also explored Venezu-elans' opinions about the "worst uses" of Venezuelan resources abroad (see Figure 3.12).[10] The first three options accounted for 61 percent of responses,

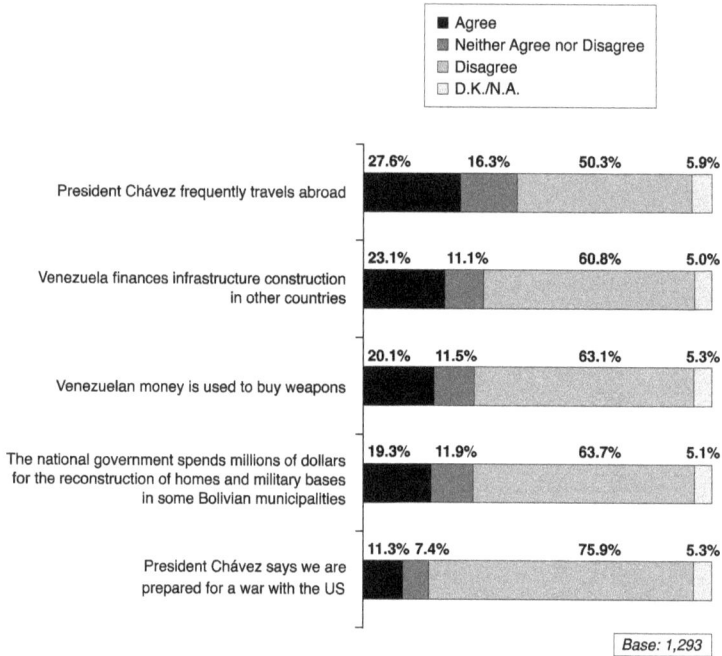

Figure 3.11. *How strongly do you agree with the following statements about Venezuelan international relations?*

1. This study was conducted among 1,293 interviewees and had the same methodological characteristics as previous DatAnalisis measurements used here (national urban coverage). Data collection was performed between July 28 and August 7, 2007. The maximum error allowed was +2.73% at the 95% confidence level.

2. Among those identifying themselves as *"chavistas"* or "pro-government" in the study, 40 percent disagreed with Venezuelan financing of "reconstruction of homes and military bases in some municipalities in Bolivia," and 47 percent agreed. The remaining percentages belonged to the "Neither agree nor disagree," "Don't know" or "No answer" categories.

indicating the high level of rejection of these contributions in Venezuelan public opinion.

More recently, the Datos polling firm measured Venezuelan public opinion about government spending abroad vis-à-vis domestic spending (see Figure 3.13).[11] The discrepancies are striking: while respondents positively evaluated those areas, such as education, health, nutrition, and housing, which they perceived as having a significant level of spending (approval levels ranging between 62 percent and 67 percent), they were much less enthusiastic about high levels of spending on initiatives related to foreign policy ("trips abroad"

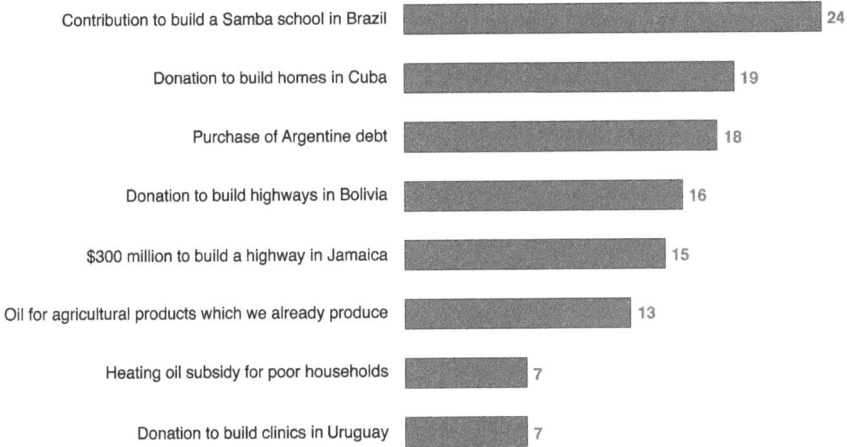

Figure 3.12. The Worst Uses of Oil Revenue (pick two)

1. Study conducted among 1,000 interviewees. The coverage included towns with more than 10,000 inhabitants and rural areas. Data collection took place between February 18 and March 5, 2006. The maximum associated error allowed was +3%.

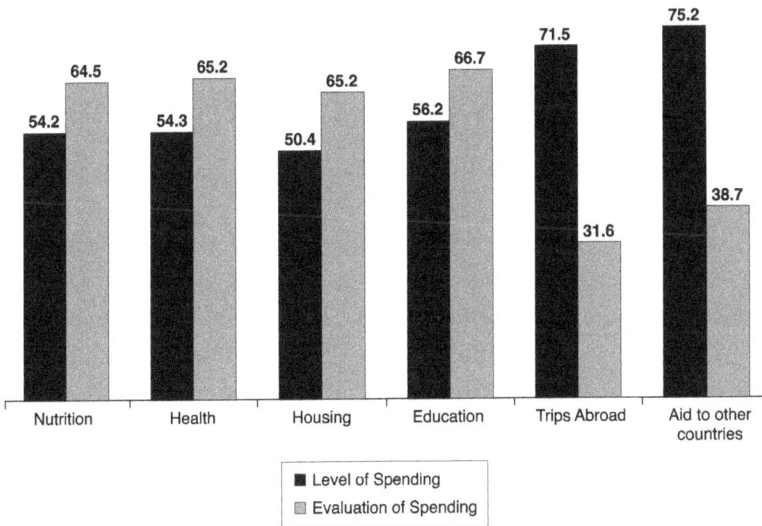

Figure 3.13. Level of Spending versus Evaluation of Spending

1. Datos's quarterly "National Pulse" poll surveyed 2,000 interviewees distributed nationally in urban and rural localities during February and March 2007.

2. Base: 2,000. Scale level of spending: Very high, High, Some, Low, Nothing. Scale evaluation of spending: Very good, Good, OK, Bad, Very bad. Only the top two are represented in the graph.

**Do you agree that the FARC and the ELN should not be considered terrorist groups?**
*Base: 2,000*

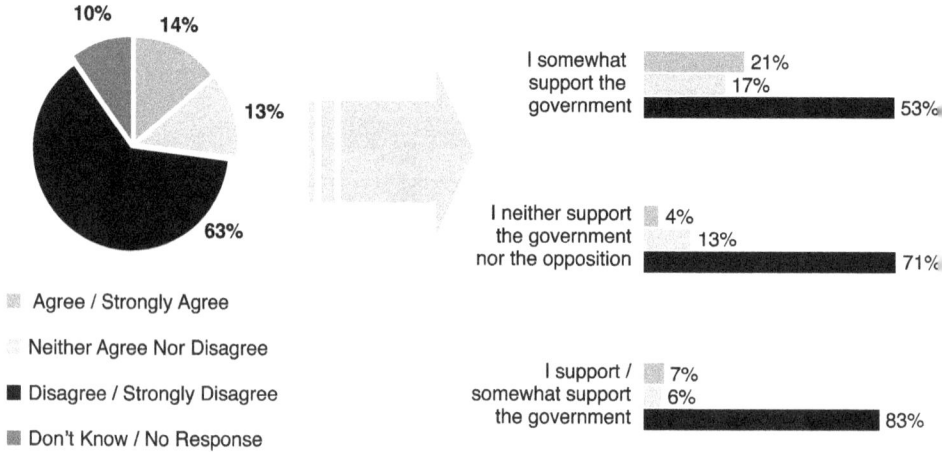

Agree / Strongly Agree

Neither Agree Nor Disagree

Disagree / Strongly Disagree

Don't Know / No Response

I support the government — 43% / 19% / 28%

I somewhat support the government — 21% / 17% / 53%

I neither support the government nor the opposition — 4% / 13% / 71%

I support / somewhat support the government — 7% / 6% / 83%

Figure 3.14. The Majority of the Population Agrees with the Classification of the FARC and ELN as Terrorists

1. Study performed during the first quarter of 2008 with 2,000 interviewees, with the same methodological design as Datos's "National Pulse" poll.

and "aid to other countries" were evaluated at 32 percent and 39 percent, respectively).

The evidence suggests that significant numbers of Venezuelans do not agree with the "assertive global projection of Venezuelan foreign policy" by the Chávez government. The reason appears to be simple, and is backed by qualitative studies: it is very difficult for Venezuelans to reconcile a foreign policy that relies so highly on spending in other countries with the poverty in their own country.[12]

Also unpopular among many Venezuelans are their president's statements and actions related to the FARC (see Figure 3.14).[13] Sixty-three percent of interviewees disagreed with a change in the status of the FARC and ELN (currently on the list of "terrorist" organizations"). Only 14 percent agreed with the proposal, with the percentage rising to 43 percent among strong government supporters (although 28 percent "disagreed," and 19 percent "neither agreed nor disagreed," meaning that 47 percent did not explicitly

**Do you agree with the position of President Chávez regarding the hostages held by the FARC?**

*Base: 2,000*

|  | Yes | No |
|---|---|---|
| 14% | | |
| 57% | | |
| 29% | | |

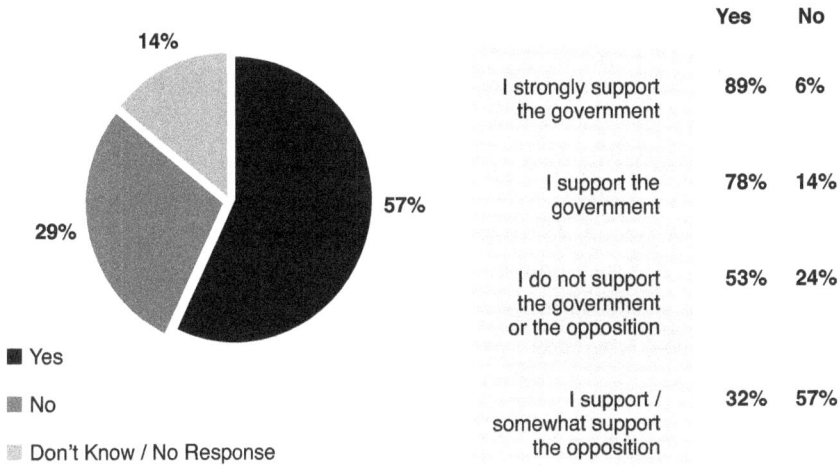

|  | Yes | No |
|---|---|---|
| I strongly support the government | 89% | 6% |
| I support the government | 78% | 14% |
| I do not support the government or the opposition | 53% | 24% |
| I support / somewhat support the opposition | 32% | 57% |

■ Yes

▨ No

▨ Don't Know / No Response

Figure 3.15. Positive public reaction to President Chávez's role in freeing the FARC hostages.

agree). Most other interviewees, including those who "somewhat" supported the government, also disagreed.

Perhaps the only positive impact of Venezuela's recent foreign policy, the same Datos poll suggests, has been Chávez's role in the release of hostages in Colombia (see Figure 3.15). Agreement for Chávez's hand in these negotiations reached 57 percent and accounted for the majority among almost all segments of the population, except for those identified as "strongly supporting the opposition," of whom 32 percent agreed and 57 percent disagreed.

## Thoughts on the Future of Venezuelan Foreign Policy

The evidence suggests that public opinion of foreign policy represents a growing challenge for the Chávez government. Perhaps the most immediate consequence will be a stronger government effort to earn a more favorable perception among key international actors, requiring the use of even more resources. Other factors suggest the same scenario: the attempted international reach of the "Bolivarian revolution"; the growing influence of the Venezuelan state in Latin America, thanks to petro-dollars; domestic challenges such as security, inflation, food shortages, a rise in social protests, and a stronger opposition; and threats to the government's image, including continued exposure of potentially incriminating information on computers owned by FARC leader Raúl Reyes and the trial in Miami of the Venezuelan accused of smuggling a suitcase containing $800 million bolívars to Argentina to support Cristina Kirchner's presidential campaign. These factors increase the pressure on the Venezuelan government to legitimate itself in the eyes of international public opinion and key actors in the inter-American community, if it plans to maintain an active foreign policy. Any such effort will increase the internal costs of foreign policy in Venezuela itself, in terms of resources and public opinion.

## Notes

1. The study was conducted among men and women over eighteen years of age in socioeconomic levels A/B, C, D, and E in Venezuela's twenty-five main urban areas. The sample size was 1,000, with a maximum allowable error of +3.04% at the 95% confidence level. The sample was collected using non-probability dynamic quotas (controlling for sex, age, and socioeconomic level). Interviews were conducted at home, face-to-face, on February 22–26, 2002.

2. The study was conducted among men and women over eighteen years of age in socioeconomic levels A/B, C, D, and E in fifty localities across the country. The sample size was 1,300, with a maximum allowable error of +2.7% at the 95% confidence level. The sample was collected using non-probability dynamic quotas (controlling for sex, age, and socioeconomic level). Interviews were conducted at home, face-to-face, on September 16–25, 2005.

3. By "anti-Americanism," we mean not only an aversion to Bush, but also an aversion to sociocultural factors; Chávez has manifested his rejection of the "American way of life" on numerous occasions, especially in his weekly TV show, *Aló, Presidente.*

4. The question was part of DatAnalysis's Omnibus Survey, with the same sample design as the one described in note 2. The Omnibus Survey had a sample size of 1,000 up to July 2003, and 1,300 thereafter, with the exceptions noted in Figure 4.

5. This content analysis was part of a larger work performed for a Venezuelan private firm from April 2004 to April 2006. All the *Aló, Presidente* programs, as well as a significant portion of Chávez's appearances on national radio and television during this period, were analyzed by the political scientist Ricardo Sucre Heredia and by the author.

6. Both studies had the same methodological characteristics, specifically, the same sampling design indicated in note 4.

7. This study used the same sampling structure and methodological characteristics as those described in note 4.

8. This study was conducted among 1,293 interviewees and had the same methodological characteristics as previous DatAnalysis measures used here (national urban coverage). Data collection was performed between July 28, 2007 and August 7, 2007. The maximum error allowed was +2.73% at the 95% confidence level.

9. Among those identifying themselves as "*chavistas*" or "pro-government" in the study, 40 percent disagreed with Venezuelan financing of "reconstruction of homes and military bases in some municipalities in Bolivia," and 47 percent agreed. The remaining percentages belonged to the "neither agree nor disagree," "don't know," and "no answer" categories.

10. The study was conducted among 1,000 interviewees. The coverage included towns with more than 10,000 inhabitants and rural areas. Data collection took place between February 18, 2006 and March 5, 2006. The maximum associated error allowed was +3%.

11. Datos's quarterly "National Pulse" poll surveyed 2,000 interviewees distributed nationally in urban and rural localities during February and March 2007.

12. According to DatAnalysis and Datos, 81 percent of Venezuelans belong to stratas D and E, the most depressed socioeconomic classes.

13. The study was performed during the first quarter of 2008 with 2,000 interviewees and had the same methodological design as Datos's "National Pulse" poll.

# 4

# Imposing the International Bolivarian Project

MARÍA TERESA ROMERO

Despite the fact that in a December 2007 referendum the majority of Venezuelans rejected President Hugo Chávez's proposal to reform the 1999 Constitution, the Venezuelan government continues its attempts to impose its national and international project, "Twenty-first Century Socialism," by any means, legal or illegal, at its disposal. The entire *chavista*-controlled state, regional, and local apparatus works furiously on every front—educational, socioeconomic, political, military, international—toward that end. The government itself has promulgated new laws and resolutions backed by a powerful political instrument, the Ley Habilitante (Delegative Law), enabling the president to rule by decree. With these powers, Chávez has been able to implement strategic actions on the domestic and foreign fronts swiftly and without consultation, working to deepen and radicalize the revolutionary process with little regard for majority opinion or consensus with principal social actors. The discussion that follows focuses on the international front, summarizing the Bolivarian government's main foreign policy initiatives to advance its domestic and international objectives.

## New International Geopolitics

The government's 2007–2013 National Economic and Development Plan, approved by the National Assembly shortly after the constitutional referendum, contains an extensive chapter on "New International Geopolitics," with the following fundamental goals:

The socialization of international relations and expansion of the socialist model abroad;

A shift to a multipolar world that challenges U.S. imperialist hegemony, and the total transformation of the United Nations and other multilateral bodies (the OAS, WTO, Mercosur, etc.);

New forms of integration to empower regions and sectors excluded under current free trade agreements that favor transnational interests, via mechanisms of political and cultural integration such as the Bolivarian Alternative for the Americas (ALBA), the Union of South American Nations (UNASUR), Petrosur, Petrocaribe, and Petroandina;

A focus on energy development as a cornerstone for the formation of a new geopolitical map;

A global strategy of political education emphasizing the achievements and objectives of the Bolivarian revolution and targeting social movements, researchers, academics, and political allies to mobilize the masses in support of the revolutionary process;

A new international communications order presenting alternative channels of information.

As can be seen from this list, Chávez's foreign policy goals are defined according to geographic zone and are based on political affinity with sympathetic governments. The list highlights two characteristics of Bolivarian diplomacy and foreign policy: expansionism and interventionism. Chávez has committed Venezuela to the strengthening of revolutionary governments and alternative movements around the world, through a strategy of "people's diplomacy" that favors cooperation with leftist sectors and social movements versus traditional contacts between governments and states.

Also to this end, the Chávez government has restructured and politicized the Ministry of Foreign Relations. As Chancellor Nicolás Maduro has noted on many occasions, the Ministry aims to ensure that every official and administrative worker in the Venezuelan Foreign Service is fully committed to the Bolivarian process and its mission abroad. In the words of Alí Rodríguez Araque, Venezuela's former ambassador to Cuba and at one time the president of Venezuela's state-owned oil company, PDVSA, "Diplomacy needs to be political because that is what an official does in a foreign country. Foreign relations are the most political of all relations." Rodríguez acknowledged that the Ministry of Foreign Relations selects the young officials most committed to the revolution, not the most prepared, to pursue Master's degrees at the Raúl Roa Institute of Foreign Relations in Cuba. A year or two after their return, according to Rodríguez, these privileged young men and women are elevated to key decision-making positions, where they are expected to effect real transformation of the Foreign Service.[1]

An unfortunate side effect of this politicization of the Foreign Service has been an unprecedented decline in professional standards. As of mid-2008, only 16 percent of Venezuela's ambassadors were career diplomats; the rest had been appointed for purely political reasons. This explains not only the discontent among many diplomatic officials, especially recent hires, but also the frequent tensions and problems in Venezuela's diplomatic missions. There have even been cases of ambassadors who have been denied *placet*—that is, they have been rejected by the country to which they are appointed—or who have been accused of meddling directly or indirectly in the host country's internal affairs. Once such instance occurred in 2006, when Argentina requested the removal of Ambassador Roger Capella after he questioned the arrest of Iranian suspects in the 1994 bombing of a Jewish cultural center in Buenos Aires. Similar cases have occurred in Colombia, Mexico, Bolivia, and Peru, all of which have accused Venezuelan ambassadors of interfering in their domestic affairs.

### Greater Rapprochement and Cooperation with Allies . . .

In pursuit of the government's international objectives, Chávez and his foreign policy team have strengthened Venezuela's commercial and political links to governments considered to be allies, both in the Americas and farther from home. Among these sympathetic regimes are some with questionable democratic reputations, most notably Iran, Syria, Russia, and Belarus.

The close relationship between Chávez and Iranian president Mahmoud Ahmadinejad has been a major source of concern in the international democratic community. The relationship has led to numerous new agreements and joint projects, the opening of a direct air route between the two countries, and an increased Iranian presence in Venezuela. Agreements have also been announced to extend bilateral cooperation in the oil sector, including the development of oil fields in Venezuela.

In Latin America, Chávez has directed commercial exchange and political and economic aid primarily to other ALBA countries: Bolivia, Cuba, Dominica, Honduras, and Nicaragua. Of all the ALBA members, Cuba has signed the most number of agreements with the Bolivarian government, for hundreds of economic and social development projects, including an oil refinery in Cienfuegos, Cuba. Venezuela has also forged numerous accords with Ecuador, Bolivia, and Nicaragua. In general, these agreements are essentially similar: Venezuela provides oil in exchange for the agricultural products that it needs to meet domestic demand. Thanks at least in part to these arrangements, the domestic and regional agendas of the presidents of Ecuador, Nicaragua, and Bolivia increasingly resemble those of their Venezuelan benefactor. Like Chávez, Rafael Correa, Daniel Ortega, and Evo Morales push constitutional reforms to increase the powers of the head of state, extend presidential terms, weaken the private media, constrain individual freedoms, and espouse nationalist and populist rhetoric over the views of an open market economy. All four presidents rely heavily on anti-American discourse and promote measures to consolidate the Bolivarian project in and outside of the Americas.

The degree of this dependency was made clear by the diplomatic impasse between Colombia and Ecuador in March 2008, when the Colombian military attacked a FARC guerrilla camp inside Ecuadorian territory. At Chávez's urging, both the Ecuadorian and Nicaraguan presidents broke diplomatic relations with Colombia. Similarly, the Bolivian government's crisis over provincial autonomy demonstrated the closeness of the relationship between Chávez and Morales. Shortly after Morales took office in 2006, the two presidents signed a complementary agreement on defense that includes military cooperation in such areas as "crisis management" and provides for the sending of military personnel in the case of "opportune national events," leaving these open to interpretation. As a result, Venezuela had an important influence over the referendums on autonomy in several Bolivian provinces in the spring of 2008. In addition to calling an extraordinary ALBA summit in Caracas to discuss the situation, and to defending Morales in international forums, Chávez repeatedly expressed his willingness to intervene directly to defend his ally.

In addition to its ideological affinity with the Morales administration, the Chávez government had other, more practical concerns: Venezuela has important oil interests in Santa Cruz and other Bolivian departments with strong pro-autonomy movements. Only days before the Santa Cruz referendum in May 2008, the state enterprises PDVSA and Yacimientos Petrolíferos Fiscales Bolivianos (YPFB) signed an agreement on exploration and drilling in the provinces of Tarija, Santa Cruz, and Chuquisaca, which are rich in hydrocarbons, and in La Paz, Beni, and Cochabamaba, which are believed to have reserves. The operation has an initial investment of $600 billion and will be run by Petroandina.

## . . . More Confrontations with the "Enemy"

Just as the Chávez government has aggressively wooed countries it considers "friends," it has pursued an increasingly confrontational, media-based campaign against governments it perceives as enemies or critics of Venezuela's domestic and foreign policies. The main targets of these diatribes have been the governments of George W. Bush in the United States and Álvaro Uribe in Colombia.

Chávez's goal in these confrontations is to provoke international reactions that will allow him to raise the nationalist flag and play the victim to supposed external aggression. This anti-imperialist, anti-Colombian rhetoric comes in handy for rallying the population around the increasingly lackluster revolutionary project and distracting public opinion from its errors, inefficiencies, corruption, and home-grown imperialism. In periods of popular discontent or before elections, conflicts with Colombia or the United States help the Chávez regime put a face on the external enemy.

Therefore, the day after Colombia's incursion against the FARC base in Ecuadorian territory, Chávez closed Venezuela's diplomatic office in Bogotá, expelled the Colombian ambassador from Caracas, and ordered the mobilization of ten battalions of the Venezuelan Armed Forces to the western border with Colombia. He condemned the Colombian action against the FARC as "cowardly murder" and called for a minute of silence in honor of slain rebel leader Raúl Reyes, whom he called a "principled revolutionary." He also repeated his recurrent calls to acknowledge the FARC as a legitimate fighting force.

The possibility of an armed conflict with Colombia, however, far from consolidating national support and boosting nationalist and anti-Colombian sentiment, provoked a strong negative reaction among the majority of Venezuelans. A survey by the polling firm Varianzas y Opinión, conducted between February 27 and March 2, 2008, found that 89 percent of Venezuelans would not support a war with Colombia. Nevertheless, Chávez has persisted with his aggressive posturing. Even after Interpol, at the request of the Colombian government, publicly presented computers belonging to Raúl Reyes detailing close financial, political, and military cooperation between high-level officials of the Chávez government and the Colombian guerrillas, Chávez merely cast aspersions and hurled insults against the governments of the United States and Colombia, and Interpol itself, an organization to which Venezuela has belonged for fifty-eight years. In frequent television and radio appearances Chávez pledged to "seriously review" political, economic, and diplomatic relations with Colombia, speculating that Uribe was capable of starting a war with Venezuela to justify U.S. intervention. He called on the Venezuelan military to be ready to fight, to perform practice missile launches, and to keep tanks and the air offensive system in operative condition.

Uribe, of course, is not the sole target of presidential insults. Chávez has also famously attacked King Juan Carlos of Spain and German Chancellor Angela Merkel, whom he compared to Hitler after she stated that the Venezuelan leader did not represent the voice of Latin America.

## Informal Support Networks and "People's Diplomacy"

In its efforts to court potential allies as well as discredit hostile powers, the Chávez government has found creative ways to pursue its own brand of social, or "people's," diplomacy, aimed at political parties, groups, movements, and social sectors ideologically akin to the revolutionary project. The mechanisms used or created for this purpose include the Bolivarian Circles abroad, the São Paulo Forum, the People's Bolivarian Congress, the Bolivarian Continental Coordinator, the Casas del ALBA, the Popular Power Regional Block, and regional alternative media, particularly Telesur.[2]

One of the most active is the Bolivarian Continental Coordinator, created by the FARC and formalized in Caracas in 2003. Its objective is to coordinate efforts across Latin America to vindicate the popular struggle against imperialism and spread Bolivarian ideals, especially among young secondary and university students, through "International Youth Brigades." During its second Congress, in Quito, Ecuador, barely two days before the Colombian military operation against the FARC in that country, the Coordinator not only declared itself a Bolivarian continental movement but also formally recognized the FARC and other insurgent groups as legitimate fighting forces.

Most alarming, it discussed the urgent need to remove President Uribe from power as soon as possible. Based on the information extracted from Raúl Reyes's computers, the choice of Quito as the venue for the Congress was not coincidental, and the Coordinator itself may be nothing more than an as yet unknown political dimension of the FARC.

The São Paulo Forum, of which the FARC has been a member for eighteen years, also aims to weaken the Uribe government, to pave the way toward a socialist revolutionary government in Colombia in the style of Venezuela, Cuba, Bolivia, Nicaragua, and Ecuador. The final declaration of a March 2008 meeting of the Forum's Working Group in Mexico City called for launching a continental offensive against the Uribe government to protest its offensive against the FARC in Ecuador.

Another of the Chávez government's foreign outreach efforts is the Casas del ALBA, which uses financing from the Venezuelan government to organize activities in favor of the Bolivarian project all across the continent. The Peruvian government in particular has issued frequent complaints about this group's activities in its territory. President Alán García has said he is "almost certain that there is foreign intervention in Peru." His defense minister, Ántero Flores-Aráoz, has described the Casas del ALBA as centers for political and ideological recruitment, visited frequently by emissaries from the FARC and officials from the Nicaraguan, Venezuelan, and Bolivian embassies, with the most activity taking place in the provinces of Puno and Ayacucho, former strongholds of the Shining Path guerrilla movement. According to Peru's Anti-Terrorism Office (Dircote), a branch of the state intelligence service, twelve members of the Peruvian chapter of the Continental Bolivarian Coordinator traveled to the group's congress in Quito with funds from the Venezuelan embassy in Lima that had been channeled through the Casas del ALBA.

## How Far Will He Go?

Chávez's aggressive and increasingly radical maneuverings have generated strong negative reactions domestically and abroad, leading to greater isolation of the Venezuelan government and an international loss of prestige. In urging U.S. approval of a free trade agreement with Colombia, President George W. Bush stressed the need to counteract the influence of Venezuela, which he described as turning into a FARC sanctuary. Documents have circulated among U.S. legislators discussing possible sanctions against Venezuela based on evidence of Chávez's links to terrorism. In Latin America, the Colombian defense minister Juan Manuel Santos has charged the Chávez government with pursuing an expansionist foreign policy, and with using its vast resources—including alleged profits from drug trafficking—to demolish democracy in the region. Alán García has disagreed with Chávez's criticism of Colombia and his argument that Colombia seeks a continental war to justify U.S. military intervention in the region. The Bolivian and Ecuadorian oppositions have organized protests against Chávez's interference in debates on provincial autonomy. Mexico's president Felipe Calderón has condemned many of Chávez's insults and statements as not contributing to the understanding or solution of problems in the region. In Europe, too, the president of the European Commission, José Manuel Durao Barroso, expressed his disapproval of the language Chávez used to refer to German chancellor Merkel, calling his words "negative statements that do not facilitate European intentions toward cooperation in Latin America." In May 2008, the day after Interpol announced its discovery of the information contained in Raúl Reyes computers, an informal meeting of the Fifth European Union Commission on Latin America and the Caribbean criticized Chávez's aggressive and war-mongering behavior.

The question is whether Chávez will find an excuse to shift from rhetoric to military action. At the Alternative People's Summit in Lima, he accused the Colombian government of illegal incursions into Venezuelan territory and brought up a century-old territorial dispute over the Guajira region along the two countries' northern border, accusing the Colombians of plans to allow a U.S. military base there. The Colombian government denied both accusations. So far, most Venezuelan analysts believe that the possibility of armed conflict between Venezuela and Colombia is remote, given economic, political, and military constraints. They view Chávez's threats and accusations as an attempt to divert attention from domestic problems and boost presidential popularity before national and regional elections. Even so, we should not lose sight of several disturbing realities.

The Chávez government has made strong and serious commitments to radical and subversive leftist groups, FARC narco-terrorists, and "rogue" states such as Iran. At the same time, it has armed itself heavily, justifying this buildup as a response to threats from Colombia and of a possible U.S. invasion. According to a recent report by the Stockholm International Peace

Research Institute (SIPRI), Venezuela was the largest importer of arms in Latin America and ninth in the world in 2007, spending $887 million on weapons, most of them from Russia, Belarus, and China.

In addition to increasing military spending, the Chávez government has reactivated thousands of militia and reservists and created a national police force to protect against alleged aggressions from the United States, Colombia, and others. The Armed Forces' 2008 budget was $4 billion in 2008,[3] but could grow even more thanks to record oil prices.

These activities raise concerns as to what the Chávez government would be willing to do in a moment of desperation to defend its Bolivarian project, whose sphere of influence extends to many foreign governments and radical leftist groups.

## Notes

1. See Valentina Oropeza, "La diplomacia tiene que ser política," *El Nacional*, May 12, 2008, 12.

2. See Lourdes Cobo, "Venezuela y el mundo transnacional: Instrumentación de la política exterior venezolana para imponer un modelo político en América Latina," in *Instituto Latinoamericano de Investigaciones Sociales (ILDIS)*, April 2008, www.ildis.org.ec/public/eventList.do

3. Castro Ocando, "Chávez amplía poder bélico," *El Nuevo Herald*, May 7, 2008.

# 5

# Venezuela's Revolutionary Foreign Policy and Colombian Security

ROMÁN D. ORTIZ

*Foreign Affairs* recently marked the sixtieth anniversary of the publication of "The Sources of Soviet Conduct," the article that became the theoretical underpinning of U.S. strategy against the Soviet Union. The article grew to mythic stature, due in part to the anonymity of its author, later revealed to be Moscow-based diplomat George Kennan, and in part to the rather mysterious way Kennan came to see the light with regard to U.S.-Soviet relations. Beyond the anecdotes surrounding its origins and its specific focus on Soviet foreign policy, however, the article's legacy lies in its attempt to overcome policy confusion when confronting an emerging power that lacks conventional foreign policy goals and does not behave in expected ways.

After nearly a decade of uneasy coexistence with the Bolivarian regime of President Hugo Chávez, Colombia has yet to produce an intellectual effort like Kennan's to try to understand the international behavior of its richest and most important neighbor. This does not mean to compare Russia under Stalin with Venezuela under Chávez; in fact, it would be difficult to imagine two scenarios more different than those of post–World War II Europe and Latin America in the first decade of the twenty-first century. Nevertheless, Kennan's writing teaches us that countries faced with the emergence of a state determined to effect radical change of the international status quo must develop new foreign policy strategies based on careful review of the new actor's motivations and objectives. In this context, two questions immediately come to mind: How is the Bolivarian regime different from other governments that have tried to influence Latin America's political orientation? And what does this mean for Colombian-Venezuelan relations?

Since Chávez took power, Venezuela's strategic influence over Colombian affairs has grown in every way imaginable. Trade between the two countries has boomed, with Venezuela accounting for $5.21 billion in goods, close to 15 percent of Colombia's exports in 2007. Chávez advocates a Bolivarian vision of Latin American unity, with a Bogotá-Caracas alliance playing a central role. His sway over Colombia's internal politics became clear in September 2007, when President Álvaro Uribe authorized him to mediate a "humanitarian exchange" with guerrilla forces that would free political and military hostages in return for the release of a number of FARC militants.

This overture ushered in the greatest crisis in the history of Colombian-Venezuelan relations, the empirical negation of Colombia's "positive commitment" to foment economic and social ties to avoid a resurrection of the two countries' historic rivalry. By now, everyone knows what happened: three times in a row, Chávez failed in his efforts to link humanitarian exchange to peace negotiations with the FARC. After each failed attempt, bilateral tensions escalated. In mid-November 2007, his direct contacts with Army Commander General Mario Montoya to discuss setting up a demilitarized zone in Pradera and Florida (Cauca) caused the Colombian government to remove him from his role as mediator, and shortly afterward Caracas announced a "freeze" on its relations with Bogotá. In late December, Chávez tried once more to serve as an intermediary in the FARC's unilateral decision to free three hostages, including a three-year-old named Emmanuel. The initiative ended in a fiasco after the Colombian government discovered the boy was already living in a Bogotá foster home. Finally, in early March 2008, Chávez announced his intention to present a peace proposal to end the Colombian conflict during the Union of South American Nations (UNASUR) summit in Cartagena later that month. His plans were cut short after the death of FARC leader Raúl Reyes in a Colombian military operation that crossed into Ecuador and sparked a regional crisis.

### Ideological Roots of the Bilateral Crisis

The deterioration of bilateral relations at a time of both spectacular growth in trade and Venezuelan peace initiatives may seem strange, but the reason isn't hard to fathom if one takes into account a factor many believed had been toppled along with the Berlin Wall: ideology. In other words, the increasingly strained relationship between Colombia and Venezuela is a practical example of the old truism that governments' foreign policies tend to reflect their domestic political projects.[1] The revolutionary nature of a regime implies a will to change not only its internal political and social order but also the international order, according to the ideology that inspires it. A government that is revolutionary at home, therefore, can also be expected to be revisionist abroad, illustrating Henry Kissinger's concept of "revolutionary potency."[2] From this perspective, the unfolding crisis in Venezuelan-Colombian bilateral relations is not the result of bad luck or personal differences between the two countries' heads of state; rather, it can best be understood as a structural conflict exacerbated by Chávez's determination to extend his revolutionary project across the continent, causing a radical shift in the strategic equilibrium.

The question is to what degree is Venezuela's international behavior determined by the ideological bases on which the regime has built and legitimated its political structures. At first glance, Chávez's rhetoric and international ventures seem like enough evidence to argue that Venezuela's foreign policy is entirely at the service of the revolution. It is difficult to look beyond the most extreme anti-imperialist discourse to explain the overtures toward Ahmadinejad in Iran, or support for Robert Mugabe in Zimbabwe. Similarly, only a determination to promote Latin American unity as a central axiom of his ideological project could justify Chávez's economic assistance to governments across the region, including $2 billion annually in subsidized oil sales to Cuba and $300 million to buy domestic support for Bolivian president Evo Morales.[3] Nevertheless, at least three arguments can be made to negate the weight of ideology in Venezuela international behavior: first, that Chávez's political message is merely the product of his eccentric personality; second, that his discourse is a cover for more pragmatic goals; and third, that his statements lack political coherency and are intended merely as propaganda.

None of these arguments disputes the revolutionary nature of Caracas's foreign policy. To begin with, Chávez's behavior, often perceived as odd or eccentric, does not cancel out his revolutionary commitment. Often, his outbursts can be interpreted as a reflection of his determination to break with the status quo, or they seem bizarre simply because the political establishment does not understand his intentions. It's enough to recall the skepticism with which many observers in the 1920s greeted Benito Mussolini's speeches from the balcony of Piazza Venezia, before Il Duce proved his ability to subvert the balance of power in the Mediterranean. The same can be said of the disdain displayed by many heads of state to the young Fidel Castro when he spoke before the U.N. General Assembly in 1960 and insisted on staying at a hotel in Harlem. Nor is a dose of pragmatism incompatible with an ideological conviction bordering on fanaticism. Examples include Lenin accepting the Kaiser's aid to overthrow Nicholas II in 1917 in exchange for an agreement for a separate peace with Germany, or, much more recently, the decision of the ayatollahs to accept Israeli weapons in the Iran-Iraq wars of the 1980s. Finally, ideological projects do not necessarily need to be coherent to mobilize or guide the political behavior of their proponents. The populist nationalism of General Velasco Alvarado, while inconsistent, was still capable of shifting the balance of power within Peruvian society and of fueling an ambitious foreign policy that brought the country to the brink of war with Chile in 1974. To take another example, the combination of socialism and Islamism in Qadaffi's "Green Book" shaped key aspects of Libya's foreign policy, including support for radical groups, from Northern Ireland to the Philippines, and the goal of Arab-African unity.

If we accept the influence of ideology on Chávez's behavior, then we must define at least the basic elements of his political project if we are to understand the forces that guide his foreign policy.[4] Even if poorly articulated, there exists a grouping of ideas which form the core of the Venezuelan leader's blueprint for "Twenty-first-century Socialism." In terms of domestic politics, the Bolivarian project is rooted in Chávez's direct relationship with the masses as head of the revolution, and with a mediating role for the Revolutionary Party to articulate popular support for the leader, and for the armed forces as the backbone of the state.[5] The alliance between the people and their *caudillo*, sustained by the party and the military, is at the heart of a political project that encompasses everything from social justice to environmental protection, crowding out legitimate interests opposed to the revolution and setting the groundwork for authoritarianism. In economic terms, the definition of the *chavista* project as socialist does not imply collectivism of the Marxist-Leninist type, but rather seems oriented toward building a state-controlled economy that does not eschew private property but intervenes in the markets to improve living standards and defend the national interest. Finally, the regime's foreign policy is linked to a commitment to Latin American unity as part of a larger Bolivarian, socialist, and anti-imperialist project.[6] This means spreading the flame of revolution across the continent to confront and defeat the United States and its allies in the region. In other words, the Bolivarian revolution is essentially a continental project that must transcend Venezuela's borders or acknowledge defeat. This combination of *caudillo*-based authoritarianism, a state-run economy, and pan-nationalism invites comparison with the Arab nationalist movement of the 1950s in countries such as Egypt, Iraq, Syria, and Libya, making it possible to look at *chavismo* as a form of Latin American "Nasserism."

## Revolutionary Foreign Policy under the Bolivarian Regime

Ideology shapes the foreign policy of a state in two main ways: it determines the government's course of action by establishing goals and the legitimate means of obtaining them; and it serves as a filter for evaluating the impact of these actions and those of other actors. Bolivarian ideology has conditioned Venezuela's foreign policy in both these ways. Venezuela has mounted a permanent foreign offensive in the years that Chávez has been in power.[7] This offensive combines elements of traditional diplomacy with a series of new tools for international cooperation, in a joint effort to export revolution to the entire continent. The results of this strategy have not always been the ones Caracas might have preferred: alongside his success in promoting friendly governments in Bolivia and Nicaragua, Chávez has suffered significant defeats in influencing elections in Mexico and Peru in 2006. Still, enormous investments of political will and financial resources have expanded Venezuela's international image and have helped make the Bolivarian regime a key actor in the region.

Chávez has pushed traditional diplomacy to the limit in promoting his revolutionary project in Latin America. He began by expanding the role of presidential diplomacy, engaging in a veritable frenzy of trips abroad. As of July 2008, the leader of the Bolivarian revolution had spent more than 450 days abroad, in a total of 150 countries, at a cost of $55 million.[8] In addition, the Foreign Relations Ministry has been expanded, its budget nearly doubling in the first two years of Chávez's presidency, from 59.1 million to 94.5 million bolívars (before the currency was reformed). Given Venezuela's confused accounting system, it isn't easy to identify exactly how the money has been spent, but in the 2008 fiscal year the Ministry's budget was 732 million bolívars—a spectacular increase in the state's foreign activities if we take into account that each new bolívar is worth ten of the old.

The increased funds at the disposal of the Bolivarian diplomatic corps have been accompanied by a sustained government effort to infuse the Foreign Service with the official state ideology. An important step in this direction came in early 2008 with a general restructuring of the Foreign Ministry. Most high-level officials were dismissed, and plans were announced to replace a substantial number of lower-level functionaries as well. The most decisive measure putting the diplomatic corps at the behest of the revolution, however, was the passage of a new Foreign Service Law. As the president of the Venezuelan National Assembly's Foreign Policy Committee, Saúl Ortega, explained, "Venezuelan diplomacy currently faces challenges that cannot be met with old-school 'cocktail diplomacy,' but rather calls for a foreign service that is prepared for a scenario of war, terrorism, and political hegemony, waged fundamentally by the United States."[9]

A series of unconventional diplomatic tools have done the most to further Caracas's ambitions on the international stage. In particular, an approach known as "social diplomacy" has channeled a limitless flow of resources to establish cooperative programs with sympathetic ideological groups. An example is the so-called "Miracle Mission," a cooperative scheme in which the Venezuelan government finances operations to help poor people across Latin America regain their sight. The initiative has had a significant humanitarian impact, saving thousands of people from blindness, but it also has a political slant: candidates for the program are selected by organizations affiliated with the Bolivarian cause. This allows groups such as the Chilean Communist Party or Mexico's Workers' Party to promote revolutionary ideas, such as a rejection of free trade, while at the same time providing free medical assistance to their supporters. The political impact of the combination is clear.

Another opportunity to extend popular support for the Bolivarian revolution is the regional integration scheme known as Petrocaribe, which guarantees Venezuelan oil at preferential rates to fifteen countries in the Caribbean Basin. Again, Caracas's generosity is indisputable; the agreement gives countries 90 days to pay 40 percent of their bill, compared to the 30 days that is the norm. The remaining 60 percent of the cost of oil is financed at between 1 and 2.5 percent interest for a period of up to twenty-five years. But here too the deal comes with ulterior motives. For starters, the arrangement has increased the beneficiaries' energy dependency on Venezuela. Caracas also encourages its clients to invest the savings on oil in social programs, and in many cases tries to influence which types of programs should be promoted and where. In supporting these social programs, the Chávez regime uses Petrocaribe resources to create popular sympathy for the Bolivarian revolution in Caribbean and Central American countries.

Caracas also supports the campaigns of political candidates across the region. Oil is often a factor in these cases as well. An example is Daniel Ortega's campaign for the Nicaraguan presidency in 2006. In this case, the Venezuelan state oil company, PDVSA, used the company Albanic (Bolivarian Alternative of Nicaragua) to supply petroleum derivatives to Sandinista mayors, who distributed them at low cost to increase popular support for Ortega. This model is apparently being repeated in El Salvador, where Venezuelan oil is channeled through ENEPASA, an association of FMLN mayors, to municipal authorities affiliated with the party.[10] Reports have also surfaced of direct payments made to candidates or political groups aligned with Caracas. The most infamous case was exposed in the trial in the United States of a Venezuelan businessman found guilty of taking $800,000 to Buenos Aires to finance the campaign of Argentine president Cristina Fernández de Kirchner. Other countries, including Peru, have complained of the proliferation of so-called Casas de ALBA that act as centers for disseminating Bolivarian ideology without clear disclosure of their funding.

Chávez has extended his generosity to fellow twenty-first-century socialist governments as a way of creating a Bolivarian axis with continental reach. In recent years, Venezuela has used the National Development Fund (FONDEN) to assume the debt of countries such as Ecuador and Argentina, giving Caracas great influence over their finances. If the selloff of its allies' bonds is any indication, Venezuela may be phasing out this model, but if we take into account the drop in Argentine bond prices after the selloff began in 2007, Venezuela's influence over its neighbors' economies is clear.

The Bolivarian regime has maintained other cooperative schemes with friendly governments across the region. Ecuador has received thousands of barrels of refined Venezuelan oil in exchange for an equivalent amount of its crude. Venezuela has also begun a massive program of cooperation with Bolivia that extends into the military sector. Former Bolivian president Jorge Quiroga has complained that the direct funding of Bolivian military units from Caracas gives the Venezuelan government enormous influence over the country's armed forces.[11]

These wide-ranging foreign ventures have had mixed results. In general, Caracas has been more or less successful in spreading its revolution, depending on the size and institutional solidity of the Latin American countries that are its target. In spite of enormous investments of political will and economic resources, Venezuela has been unable to decisively influence elections in countries such as Mexico or Peru. Its efforts have met with more success, however, in the small and fragile democracies in Central America and the Caribbean. There, Caracas seems to have won considerable influence, which could spread even farther in the not so distant future.

## The Ideological Breach in Colombian-Venezuelan Relations

At the same time, the revolutionary orientation of the Bolivarian regime's foreign policy has opened a profound breach in its relations with Bogotá. The governments of Venezuela and Colombia may agree on increasing bilateral trade or humanitarian accords, but the strategic objectives behind these goals are very different. For the Colombian government, for example, increasing its exports is a strictly economic goal, important for stimulating nontraditional productive sectors. The Chávez regime's commitment to socialism, however, makes it impossible for it to view trade in the same way. For the Venezuelan government, trade is a strategic tool that furthers integration between the two countries while offering influence over Colombia's domestic politics through pressure on economic interests that depend on the Venezuelan market. The two governments' positions during successive bilateral crises offer ample evidence of their diverging perceptions of the value of trade. While Bogotá tries systematically to shield trade relations from the ups and downs of politics, Caracas just as systematically uses trade as a political tool, despite the potential consequences for the Venezuelan as well as Colombian economies.

A similar dynamic holds true for the idea of humanitarian exchange. Both presidents share a common concern for the fate of the FARC's hostages, but the convergence ends there. For the Uribe administration, an accord to free kidnap victims is a purely humanitarian matter that does not imply the beginning of a peace process with the guerrillas. For Chávez, in contrast, it represents an opportunity to promote negotiations with the FARC that condition an end to violence on the radical modification of Colombia's political and economic system according to Bolivarian precepts. This process would not only impose "twenty-first-century Socialism" on Colombia, it would also consolidate Venezuela's position as the capital of the project for continental unity. This vision of "humanitarian exchange" as a practical tool at the service of an ideology should come as no surprise if one considers the ideological closeness between the FARC and the revolutionary regime in Caracas. The guerrillas have declared their sympathy toward Chávez and his socialist project on numerous occasions, and Chávez in turn has described the FARC as "insurgent forces with a political project, a Bolivarian project, that is respected here." This affinity explains Chávez's determination to have the FARC accepted as a legitimate fighting force, to bring it closer to obtaining political concessions that would open the door to the extension of the Bolivarian project to Colombia.

If any doubt remained of the ideological nature of the breach that separates the foreign policies of Bogotá and Caracas, it was dispelled after the discovery of proof of the cooperation between the Chávez government and the FARC. According to information contained in computers belonging to Raúl Reyes, which was captured during the raid on his camp, the Bolivarian regime's collaboration with the guerrillas extends beyond supporting their efforts to gain international political recognition, to sending them money and arms and allowing them to use Venezuelan territory for their operations against the Colombian military.[12] This strategic alliance implies agreement between the Venezuelan government and the FARC as to the illegitimate nature of the Colombian political regime and the need to resort to violence to destroy it, replacing it with an alternative model defined in general terms as nationalistic, anti-imperialist and populist. This will only be accomplished through head-on clashes with the government and the overwhelming majority of Colombian society, which views the FARC as a terrorist group lacking any political justification, and which supports the state military campaign defining it as an existential threat against democracy.

The deterioration of Colombian-Venezuelan relations, therefore, is the result of the ideological clash between the political projects embodied by the two countries' leaders.[13] The underlying causes of the recent crises in bilateral relations are not circumstantial factors, such as the personalities of decision-makers or diplomatic gaffes committed by one or the other party. Such factors could undoubtedly help explain why tensions may have erupted at a certain moment in time, or why an ongoing crisis suddenly worsened. Beneath these more superficial concerns, however, the basic logic of successive bilateral crises is a result of the inevitable clash between Caracas's desire to extend its revolutionary project as the only way to realize its dreams of "Bolivarian unity" and Bogotá's resistance to this plan. This conclusion has important strategic implications for the future of bilateral relations. It means that current tensions are not merely a passing phase, but rather they are a signal that the nature of relations between the two countries has experienced a structural change. Tensions may abate from time to time, as a consequence of dialogue or negotiations, but such rapprochements are likely to be temporary, due to the ideological antagonism that underlies Venezuelan-Colombian relations.

The status quo seems set to continue as long as the Bolivarian regime views extending its revolution as the guiding principle of its foreign policy. The Chávez government could eventually abandon its goal of continental unity, but such a change will not come easily. Even before he took power, Chávez had based his legitimacy and social support on the idea of a continental movement to create a united and socialist Latin America. Renouncing the extension of the revolution, therefore, would be the same as admitting the impracticality of the Bolivarian project. The implications would go beyond a foreign policy defeat to affect the very foundations of the regime and would call into question its survival. In other words, for the Chávez regime, foreign policy and internal legitimacy are indissolubly linked: changing the first inevitably weakens the second. Only under exceptional circumstances, such as an economic or institutional crisis, is it possible to imagine the Chávez government assuming the political costs of renouncing the Bolivarian project. Venezuela's current foreign policy, therefore, seems likely to endure for as long as the Bolivarian regime itself.

## Can the Revolution Survive?

The Bolivarian revolution is not experiencing its best moments. The victory of the "No" vote in the December 2007 referendum on turning Venezuela into a socialist state was Chávez's first defeat in nine consecutive elections.[14] It was followed by significant economic problems. Venezuela closed 2008 with 22.5 percent inflation and shortages of basic consumer products, an especially serious scenario considering the government's efforts to protect the economic interests of the popular sectors that are its base of support. These political and economic setbacks do not seem critical enough to produce the regime's downfall in the short term, however; Chávez may have lost the referendum, but 4.5 million people (49.3 percent) voted in his favor, and many of those who abstained probably would have voted "Yes" if his continuity in power had been at stake. Venezuela's enormous oil revenues provide a cushion against hyperinflation, at least in the short term, and the government's vast purchasing power has allowed it to compensate for domestic shortages with massive imports of consumer products.

This does not mean that Chávez is guaranteed to stay in office indefinitely. Ultimately, the regime's future depends on the evolution of its income from oil. Two uncertainties affect this evolution: on the one hand, the volatility of oil prices; and, on the other, the deterioration of infrastructure in the energy sector. Venezuela's capacity to extract crude has decreased by 30 percent since Chávez took power, the result of a lack of investment and deficient performance.[15] Neither of these two factors has reached the point of threatening a short-term economic collapse, and even if such a collapse were to occur, Chávez has enough resources to deliver massive social benefits to conserve political support. This strategy could already be seen in the months preceding the November 2008 state and local elections, when the government announced a 30 percent salary increase for public employees. The Bolivarian regime could suffer a gradual loss of political prestige as the public grows weary of inefficient government performance and corruption, but as long as oil prices remain high, Chávez can maintain social spending at a level that guarantees the support of a significant portion of the Venezuelan public.

Under these circumstances, the Bolivarian regime could endure long enough to oblige Colombia to rethink its foreign policy toward its most important neighbor. The inevitable impact on Colombia's internal conflict makes the need for such a change more urgent. The Uribe administration's military offensives have pushed illegal armed groups away from wealthy and populous regions to border zones where the state's actions are complicated by hostile terrain and the insurgents' ability to seek refuge in neighboring countries. Raúl Reyes's camp was only a little more than a mile inside Ecuador, a sign of the strategic importance of foreign borders to guerrilla and criminal organizations. Control over its borders is likely to become one of Colombia's key challenges in completing the pacification of its territory. At the same time, Chávez has pushed to award the FARC political recognition, and his strategic objective is to create the conditions for ending the Colombian conflict through negotiation rather than a definitive military victory.

The ideological breach between the two governments makes it impossible for Bogotá to count on Caracas for cooperation in the battle against terrorism, and Colombia needs a security strategy that takes this reality into account.

The challenge for Colombia will become more complex as Chávez develops a network of alliances across the region. This was the main lesson of the recent crisis following Colombia's incursion into Ecuadorian territory, when Venezuela quickly became the leader of an anti-Colombian front, along with Ecuador and Nicaragua. The three governments had different motives for escalating their hostility toward Colombia: Venezuela's main objective was expansion of its "twenty-first-century Socialism" project; Ecuador sought the political expediency of nationalism to shore up a fragile domestic base; and Nicaragua saw an opportunity to support its side in a dispute with Colombia over maritime borders while diverting attention from problems at home. The Caracas-Quito-Managua axis is hardly a united front, but while disagreements may arise among the region's Bolivarian governments, it would be a mistake to underestimate the political ties linking Chávez, Correa, and Ortega. They may be motivated by different goals and united by purely pragmatic interests, but the three have nationalism and the anti-liberalism of the Bolivarian message in common. The ideological breach that separates Bogotá from Caracas extends to other capitals in the region, contributing to a certain isolation of Colombia.

## Conclusions: A New Security Policy for Colombia?

Bogotá faces a new and challenging regional scenario that could put at risk many of its gains in domestic stability and erode its position in the Latin American arena. Responding to this challenge calls for an intellectual effort to reformulate Colombia's foreign policy based on the understanding that the strategic environment has changed and requires a review of the state's lines of action. A good place to start would be the assumption that the ideological breach with Caracas is wide enough to call into question the viability of regional integration. In fact, Colombia should prepare for a structural increase in the conflictive nature of its relations with Venezuela. Acceptance of this fact is the necessary starting point to allow for the successful management of future tensions and acceptable levels of cooperation given the differences that separate the two governments.

Building on this foundation, Colombia should study and, if necessary, implement measures in three areas. First, it should reconsider its security policy given the reality of new international challenges. This effort should be oriented toward giving the armed forces a basic level of deterrence, to avoid making the state vulnerable to threats from neighboring countries with overwhelmingly superior military might. At the same time, measures to increase mutual trust among the region's governments should be pursued to stabilize the balance of military power in the Andes. Second, Bogotá should reevaluate its commercial relations with neighboring countries. No one can deny the need to maintain the flow of trade with Venezuela, not only because of the economic benefits it brings but also because of its role in stabilizing bilateral relations. Nevertheless, finding alternative markets should also be a priority, to reduce the susceptibility of Colombia's balance of trade to swings in the Venezuelan economy, which has proven itself vulnerable to political developments. In this sense, it is impossible to overemphasize the importance of free trade with the United States. As far as Venezuela is concerned, Colombia must pursue international juridical and political mechanisms to defend its investments against possible attempts at nationalization. Finally, Colombia should modify its foreign policy with the aim of ending its isolation in the region. Beyond the Venezuela-Ecuador-Nicaragua alliance, countries such as Peru, Chile, or Mexico have political positions much closer to Colombia's. The problem is not so much a lack of allies in the region, but rather the lack of a strategy to articulate a political counterweight to the emergence of a "Bolivarian axis." Political, trade, and security relations with the United States should also be strengthened. Closer ties with Washington do not contradict improving relations with other Latin American countries; in fact, the two are complementary. Colombia could actually gain prominence among its Latin American neighbors if it can develop a privileged relationship with the United States and serve as a bridge between North and South.

## Notes

1. Particularly interesting in this sense is the relationship between the autocratic nature of certain political regimes and a foreign policy that goes against the status quo, as Carl J. Friedrich and Zbigniew K. Brzezinski describe in *Totalitarian Dictatorship and Autocracy* (Cambridge, Mass.: Harvard University Press, 1965,) 353.

2. Henry Kissinger, *Un mundo restaurado* (Mexico City: Fondo de Cultura Económica, 1973).

3. The Venezuelan government has doled out massive sums in foreign aid to allied gov-

ernments in Latin America and elsewhere, but the exact amount of money invested is uncertain. Some statistics can be found in Gustavo Coronel, "The Corruption of Democracy in Venezuela," PetroleumWorld.com, March 9, 2008, www.petroleumworld.com/sf08030901.htm.

4. For a more extensive analysis of the ideology of Venezuela's Bolivarian regime, see Román D. Ortiz, "Venezuela: Una revolución en crisis," *Cuadernos Hispanoamericanos,* no. 622 (April 2002): 87–98; and the collection of Chávez's speeches in Agustín Blanco Muñoz, *Habla el Comandante* (Caracas: Cátedra "Pio Tamayo"–CEHA/IIES/FACES/UCV, 1998).

5. Although the Venezuelan president has denied any political influence from the ideas of Norberto Ceresole, the writings of this radical Argentine ideologue were long considered to be one of the theoretical foundations of the Bolivarian revolution. Some of Chávez's key political ideas, such as the central role of the military, can be found in Ceresole's works (see Alberto Garrido, *Mi amigo Chávez: Conversaciones con Norberto Ceresole* [Caracas: Ediciones del Autor, 2001], especially p. 115).

6. For more details about the foundations of the Bolivarian regime's foreign and defense policies, see Alberto Garrido, *Chávez: Plan andino y guerra asimétrica* (Bogotá: Intermedio, 2006), especially pp. 19 and 79.

7. A summary of Chávez's international offensive during his first years in power can be found in Cristina Marcano and Alberto Barrera Tyszka, *Hugo Chávez sin uniforme: Una historia personal* (Caracas: Debate, 2005), 283.

8. "Viajes del mandatario suman 450 días," *El Universal* (Caracas), July 19, 2008, http://eluniversal.com/2008/07/19/pol_art_viajes-del-mandatari_953087.shtml.

9. "Diplomacia en la mira," *El Universal* (Caracas), January 30, 2008, http://buscador.eluniversal.com/2008/01/30/pol_art_diplomacia-en-la-mir_691792.shtml.

10. "Edmundo Jarquín acusa a Chávez de apoyar a Ortega con petróleo barato," *El Nuevo Diario* (Managua), October 10, 2006, http://impreso.elnuevodiario.com.ni/2006/10/10/politica/31021; "Chávez firma acuerdo de energía con El Salvador," *La Prensa Gráfica* (San Salvador), March 21, 2006, http://archive.laprensa.com.sv/20060321/lodeldia/286.asp; "Alba Petróleos tiene fondos sin respaldo," *La Prensa Gráfica* (San Salvador), August 14, 2008, www.laprensagrafica.com/lodeldia/20080814/17171.asp.

11. "Jorge Quiroga: Morales es un títere de Venezuela," *El Universal* (Caracas), March 26, 2008, http://buscador.eluniversal.com/2008/03/26/int_ava_jorge-quiroga:-moral_26A1461519.shtml.

12. See Maite Rico, "Los papeles de las FARC acusan a Chávez," *El País* (Madrid), May 10, 2008; and "Los e-mails secretos," *Semana* (Bogotá), May 19–26, 2008.

13. Some of the main architects of the Bolivarian republic's defense policy consider Colombia to be a threat to the regime on the basis of ideology and its close ties to the United States (see Alberto Muller Rojas, "El problema binacional de la seguridad estratégica," in *Colombia-Venezuela: Retos de una convivencia*, ed. Socorro Ramírez and José María Cadenas [Bogotá: Central University of Venezuela, IEPRI-Center for América Studies, 2007], 377).

14. For an analysis of the referendum, see Ana María San Juan, "Referéndum del 2D en

Venezuela: Balance y perspectivas," December 2007, www.cartercenter.org/resources/pdfs/Ana_Maria_Analisis_2D_en_Venezuela.pdf.

15. Joe Carroll, "Oil Output Undermined by Chávez, CERA Says," Bloomberg.com, February 12, 2008, http://www.bloomberg.com/apps/news?pid=20601086&refer=news&sid=aR1myfZNWueI.

# 6

## Strangest Bedfellows

The Belarus-Venezuela Connection

RALPH S. CLEM

William Safire, known for his etymological genius as well as for his incisive conservative commentary, provided us with both the background and an excellent definition of the much-used phrase, "Politics makes for strange bedfellows." As is so often the case, Safire tells us, the expression traces back to Shakespeare (in this case, to *The Tempest*), but its present-day usage is defined as "members of an unlikely alliance, often attacked as an 'unholy alliance,' forced by circumstances to work together."[1] The subject of this discussion—the increasingly close friendship between the Bolivarian Republic of Venezuela and the Republic of Belarus (formerly, in English, Belorussia)—is about as unlikely an alliance as one might find, and is, in the minds of many, also unholy.

Venezuelan president Hugo Chávez ardently pursues contacts and alliances with a variety of countries, some out of ideology and others as a matter of convenience. Russia has made huge multi-billion dollar arms sales to Venezuela; China, in its relentless pursuit of hydrocarbons, is a growing presence in the country; and Iran finds a kindred spirit in Chávez's anti-U.S. rhetoric. These relationships, involving major players on the international stage, are fairly straightforward and not difficult to understand. But Venezuela and Belarus? Why have these two countries, separated by great distance and very different cultures, become so closely linked, to a degree that early skeptics certainly misjudged? A related question is, Are these ties meaningful, in the sense that they satisfy some genuine needs, or is the relationship ephemeral and without real substance?

To frame these questions, the discussion that follows focuses on three topics: regime legitimacy, economic cooperation, and military assistance. As will be clear below, the second topic is an important element of the first, and the third is actually a subset of the second, but because of the high visibility and sensitivity of foreign military involvement in Latin America, it will be given special attention here. The emphasis will be on the Belarusian side of the association, as so much expertise is available in this volume on Venezuela. Given that relatively little is known in general about Belarus, some general background on the country is included, especially its recent (i.e., post-Soviet) political and economic situation.

## Belarus: Legitimizing an Accidental State

If, for whatever reason, one decided to fly from Caracas to Minsk, the capital of Belarus, the journey would traverse some 5,820 nautical miles (9,365 kilometers) as the crow flies; or, as the airlines fly (according to Expedia.com), it would take a minimum of 16 hours and 25 minutes, with two intermediate stops required. The trip would also, in many respects, take one back in time. Belarus presently can be described as a Stalinist state, arguably "the land where the Soviet Union never really went away," and it is often referred to as "Europe's last dictatorship."[2] Belarus today is about the size of Kansas, with a population of around 9.7 million. As a glance at the map will reveal, the country is wedged into the historically hazardous territory between Poland and Russia, a region that, unfortunately for its inhabitants, is astride the corridor through which invading armies have flowed at least since the time of the Mongols. Napoleon and Hitler came this way, moving east; the Russian Imperial Army and later the Red Army pushed them back to the west. Controlled by the medieval Lithuanian Kingdom, then by Poland, the Belarusian lands were annexed by Russia in the late seventeenth century and became part of the USSR after the transition to Soviet rule.[3] Sometime around the turn of the twentieth century, at least a modicum of Belarusian ethnic consciousness was coalescing. Belarusians are a Slavic people, close linguistically to the Russians to the east and, less so, the Ukrainians to the south.[4] This ethnic distinctiveness was sufficient to warrant the creation of a Belorussian Soviet Socialist Republic, as part of the Soviet federation, from 1922 until the dissolution of the USSR in 1991, after which the Republic of Belarus became an independent state.

After independence, Belarus flirted briefly with democracy until 1994, when Aleksandr Lukashenka (sometimes written as Lukashenko) was elected president. In 1996, Lukashenka engineered a referendum that greatly enhanced the power of the presidency and extended his term in office to 2001, when he was reelected. In 2004, another referendum abolished term limits, allowing Lukashenka to run and win once again in 2006. None of these elections or referenda, or any others to the National Assembly or regional offices, has been judged free and fair by international monitoring bodies. With a fully compliant legislature and a massive and highly intrusive secret police apparatus (the only remaining KGB), the fifty-six-year-old Lukashenka is unlikely to leave office anytime soon.

The inexorable erosion of human rights in Belarus has resulted in the country's becoming a virtual pariah state. Condemned by every non-governmental human rights organization, the Lukashenka regime has murdered political opponents, imprisoned opposition party members, reinstituted forced labor camps and internal exile, and brutally suppressed peaceful marches with truncheon-wielding riot police. The United States and the European Union have imposed economic sanctions, including asset seizures, against Belarus, and have placed travel restrictions on Lukashenka and high-ranking members of the government and security forces. In response, Belarus ordered the U.S. ambassador to leave the country; she departed in March, 2008, and as of this writing both sides seem ready to close down their respective embassies.[5] Further compounding Belarus's isolation is the deterioration of relations with Russia. For more than a decade, the two countries courted each other over the idea of a possible union, if not a merger. This dance came to an abrupt end in late 2006 and early 2007, when Russia unilaterally raised the cost of its natural gas and crude oil exports to Belarus, prompting a threat from the latter to close the pipelines that transit the country and carry a significant share of Russia's energy exports to Europe. A long-term supply disruption was avoided, but the event forced Belarus to adopt a more independent stance vis-à-vis Russia.[6]

As Belarus became ever more estranged from its neighbors, Lukashenka required other international ties to legitimize his status and that of his regime. Interestingly, this "quest for . . . international prestige," as David Marples puts it, derives in large measure from the explicit rejection of the more traditional ethnic-nationalist raison d'être that has characterized, to varying degrees, all of the other post-Soviet states. Even though more than 80 percent of the country's population is ethnically Belarusian, and even though Soviet policies promoted the strengthening of Belarusian ethnicity, traditional Belarusian national symbols, including the white-red-white flag, have been banned from public display and Russian remains the dominant language. This may be because the Belarusian ethnic identity was not as mature as that of Russia, Armenia, or Estonia, for example. Absent the emotional power of nationalism, the Lukashenka regime had to search for another basis of state legitimacy.[7]

Consequently, Belarus strives to portray itself as a fully participating member of the international community and regional organizations. It has engaged actively in bilateral relations with countries across the globe, reinforcing the image of a legitimate state by acting like a state. Although geography dictates a foreign policy focus on Russia and Europe, Belarus's growing ties with Latin America serve, among other things, to bolster its image as a viable and sustainable state, and to further the aims of the Belarusian government insofar as stability and economic growth are concerned.[8] Within this expanding Latin American presence, the relationship with Venezuela is clearly the top priority, and the personal ties between Lukashenka and Chávez appear both cordial and important to each of them.

To develop this new relationship, Chávez made two state visits to Minsk (July 2006 and June 2007), followed by a trip by Lukashenka to Caracas (December 2007), all of which received extensive coverage in the Belarusian press. At the 2006 meeting in Minsk, Chávez used the term "strategic alliance" to describe the budding friendship, with efforts to "counter U.S. imperialism" as one of its main goals.[9] Upon arrival, Chávez told reporters that "Belarus is a model of a social state, which we are also building. We must defend the interests of the individual and not the hegemonic interests of the capitalists, wherever they may be, in Europe or Latin America."[10] During both the 2006 and 2007 Minsk meetings, Lukashenka echoed the "strategic alliance" theme, raising the possibility of greater cooperation between the two countries in a number of economic and military areas, "all of which will lead to increasing prosperity of the Venezuelan and Belarusian peoples."[11] Playing up the theme of victimization by hostile forces, Lukashenka added, "It is quite evident and clear at whose bidding mass media are acting in this case. But today I have met a person [Chávez] who has brilliantly mastered the issues of defense and security, who has a most profound knowledge of his country's economy and who whole-heartedly loves his nation."[12]

Beyond the pomp and circumstance, the reviewing of honor guards, and meetings with schoolchildren and workers, not to mention Lukashenka laying a wreath at the tomb of Simón Bolívar—what does it all mean? Although to some extent symbolic, symbolism matters, and for a leader and a country seeking validation, the many photo opportunities that such high-level visits present are an important source of regime legitimacy. The two leaders signed bilateral agreements during both of Chávez's visits; if, as we will examine below, these agreements actually amount to something substantial and deliver on the promise of greater prosperity, then that prosperity in turn will reinforce their claims to benefit their respective peoples.

### Economic Cooperation: Promise or Reality?

Public opinion polling in Belarus reveals that the majority of its citizens rank economic security as their highest priority.[13] Consequently, the extent to which Belarusians believe that Lukashenka's policies have succeeded in growing their country's economy is a key element, perhaps *the* key element, in regime legitimacy. Certainly, the Lukashenka government takes credit for achieving economic stability, and this may indeed be its appeal to those who support, or at least acquiesce, to the regime. Despite some dispute over the sources of economic growth and the "true" numbers to measure it, there is at least general agreement that under Lukashenka the Belarusian economy has managed a respectable level of sustained growth for about a decade, avoiding some of the worst consequences of the post-Socialist period as experienced by other former Soviet states.[14] Belarus has also maintained a high level (around 80 percent) of state ownership of the means of production, re-nationalized previously privatized firms (including banks), and through redistributive measures insured a high degree of income equality.[15] For our purposes, however, the question is whether or not the economic cooperation agreements with Venezuela will in fact bear fruit and contribute to further economic growth in Belarus, and, by extension, to the continuation of the Lukashenka government.

The Venezuela-Belarus High-Level Bilateral Committee, established after Chávez's first visit to Minsk, undertook to "exchange proposals in the areas of energy, oil and petrochemicals, technology and science, technical and military cooperation, housing, habitat and infrastructure, agriculture, machinery and food."[16] Given the structure of Venezuela's economy, especially its export sector, it is no surprise that one of the specific areas in which the two countries propose to cooperate is the petroleum industry. Although Belarus has few indigenous energy resources, the country does have major refining capacity and a large petrochemical industry, with main installations at the NAFTAN refinery in Novopolotsk and the NPZ in Mozyr. Of course, these are nothing close to the capacity of Venezuela's state-owned Petróleos de Venezuela, S.A. (PDVSA). In 2007, PDVSA and the Belarusian state oil company, Belarusneft, commenced cooperative ventures, giving the latter drilling rights in the Junin I block. The joint venture Petrolera BeloVenesolana (PDVSA and Belarusneft) announced plans to extract some 900,000 tons of oil in 2008, from which Belarus would derive 40 percent of the profits—around US$300 million at May 2008 prices.[17] The Belarusian state petrochemical company, Belneftekhim, has also said it will refine oil in Venezuela and market products in Latin America.[18] The profits from these ventures, which are significant given the relatively small size of the Belarusian economy, are in effect a subsidy from Venezuela to Belarus.

Belarus also has a strong interest in Venezuela as a market for manufactured goods and as an entry point into the broader Latin American market. As Lukashenka stated just prior to his trip to Caracas, "The importance of the Venezuelan vector of our foreign policy lies, first of all, in the fact that it allows Belarus to consolidate its position in the Latin American Region."[19] That said, the question is: What does Belarus make that Venezuela might want? The most obvious answer is tractors, buses, trucks, and other vehicles. As a legacy of the Soviet period, Belarus is home to several huge vehicle manufacturing plants. The giant MAZ combine (Minskiy Avtomobilniy Zavod, or Minsk Automobile Factory), the Minsk Tractor Factory (MTZ), and other firms have exported their wares to Venezuela and have plans to build assembly plants there for both local and regional markets. One potential dark side to this story is the sudden shortage of powdered milk and baby formula in Belarus, a crisis whose origin seems to be in the state's decision to export these products to Venezuela, in what we might call an "oil-for-formula" deal.[20]

What is in this for Venezuela in economic terms? The simplest answer is that Chávez can count on Belarus to assist in his Bolivarian industrial revolution. Joint ventures in manufacturing using Belarusian technical expertise, along with that of other countries sympathetic to the Bolivarian cause (Russia, Iran, Brazil), could allow for substantial industrial growth in Venezuela in coming years. The relationship is symbiotic, giving Venezuela access to both foreign technology and investment funds, neither of which that country finds it easy to obtain otherwise.

## Belarusian Military Assistance to Venezuela

Like most former Soviet countries, Belarus inherited a substantial stockpile of military equipment and infrastructure and, perhaps more important, an extensive military technology and manufacturing base. As its part of the spoils when the USSR collapsed, Belarus acquired 378 aircraft, 3,100 tanks, and 3,400 other armored vehicles, much more equipment than its new armed forces required. The Belarusian political and military leadership was quick to realize that much of this weaponry and military hardware could be sold in the global arms trade, and soon everything from sophisticated aircraft to Kalashnikov assault rifles was on its way to Syria, Iraq, Yemen, Cuba, Ethiopia, and Libya, among other destinations. Eventually, of course, Belarus used up its supply of ex-Soviet military equipment, and so changed its sales strategy to focus on refurbishing or upgrading weapons systems of Soviet manufacture, for which activities it had an excellent industrial infrastructure. After Lukashenka came to power, much greater emphasis was placed on advanced military technology sales: avionics, radars, electro-optical devices, and fire control systems became big ticket export items. All of this effort, which ultimately lifted Belarus to top-ten status among the world's arms exporters, is coordinated by the State Military Industrial Committee (Goskomvoenprom), which reports directly to Lukashenka.[21]

Most sales were to the "usual suspects"—that is, countries that had formerly been equipped with Soviet military hardware. More recently, however, Belarus's customer list has expanded to include, most importantly for our purposes, Peru, its first major Latin American client. In 1996, Belarus sold the country a squadron of MiG-29 fighters, another squadron of Su-25 ground attack aircraft, the "Nebo" radar system, and the "Igla" surface-to-air missile system, followed in 1998 by thirty-five multiple launcher rockets systems. According to various news reports, the aircraft sales in particular were problematic for Peru; they proved difficult to maintain, and Belarus did not provide much in the way of customer service.[22] In 2007, Belarus made its first important sale to Venezuela: night vision scopes for assault riles, an item that Belarus specializes in and sells for considerably less than U.S. or European models.[23] More significantly, in April 2008, Belarus announced it would design and build a national integrated air defense system for Venezuela. This again plays to the strength of Belarus's military-industrial sector, which has long-standing expertise in such systems.[24] Although no cost estimate is publicly available, this multi-year project elevates military cooperation between the two countries, certainly not to the level of Russia's defense industries, which have benefited tremendously from sales to Venezuela, but still significantly, given the upgrade it represents for Venezuela's defense capability. Again, in this area there is symbiosis for Belarus, which needs the business, and Venezuela, which must circumvent the U.S. arms embargo imposed on it in 2006.

One postscript to this discussion involves the possible ties between Belarusian arms dealers, the Chávez government, and the FARC (Fuerzas Armadas Revolucionarias de Colombia) guerrilla movement in neighboring Colombia. According to media reports, a laptop computer seized by Colombian commandos in a cross-border raid against a FARC camp in Ecuador contained references to small arms shipments from or involving Belarus.[25] Belarus has officially denied the allegations, and the ultimate resolution of the issue remains unclear.

## Conclusions

The seemingly counterintuitive strategic alliance between Belarus and Venezuela is, on closer examination, of value to both sides for several reasons. In each case, high-visibility state visits, replete with ceremonial occasions and extensive press coverage, lend legitimacy to the respective heads of state, and thereby to their governments. These events allow for mutually reinforcing statements of solidarity and the opportunity to vilify hostile governments, especially that of the United States, further legitimizing the regime by invoking an "us-versus-them" threat. As a practical matter, both sides potentially benefit from wide-ranging economic cooperation, although it is too early to judge the success of such efforts. Ties with Venezuela will likely provide additional markets for Belarusian manufacturing firms and, if the petroleum joint ventures succeed, a significant infusion of cash as well. Venezuela stands to gain from technology transfers and licensed manufacturing. Finally, given the importance of the arms industry to Belarus, Venezuela is an important new market, even if the level of sales and services is presently small. For Venezuela, given its hypernational defense posture, access to another willing top-ten military supplier is reassuring, especially one with considerable expertise and capability in advanced technology. As long as the Chávez and Lukashenka governments remain in power, therefore, we can expect the Belarus-Venezuela relationship to flourish, and that relationship in itself reinforces both regimes.

## Notes

1. William Safire, *Safire's New Political Dictionary* (New York: Random House, 1993), 762–63.

2. Nick Paton Walsh, "Europe's Last Dictatorship," *The Guardian*, March 2, 2006.

3. The western areas of what is today Belarus were held by Poland in the period between the two world wars.

4. Ralph S. Clem, "The Belorussians: The Formation and Maintenance of National Identity," in *The Nationalities in Gorbachev's Russia*, ed. Graham Smith (London: Longman's, 1990, 1995), 142–60.

5. "Belarus Expels U.S. Ambassador," *International Herald Tribune*, March 7, 2008.

6. David R. Marples, "Elections and Nation-Building in Belarus: A Comment on Ioffe," *Eurasian Geography and Economics* 48, no. 1 (January–February 2007): 60–61.

7. Clem, "The Belorussians"; Grigory Ioffe, "Unfinished Nation-Building in Belarus and the 2006 Presidential Election," *Eurasian Geography and Economics* 48, no. 1 (January–February 2007): 37-58.

8. Belarus Ministry of Foreign Affairs, "Cooperation between Belarus and Latin American Countries," www.mfa.gov.by.

9. Claire Bigg, "Russia: Chávez to Seal Arms Deal," *RFE/RL News and Analysis*, July 25, 2006, at www.rferl.org.

10. "Chávez Praises Belarus," *New York Times* (from Reuters), July 24, 2006. Also see Steven Mather, "Venezuela and Belarus Forge 'Strategic Alliance,'" Venezuelanalysis.com, July 25, 2006, at www.venezuelanalysis.com/.

11. Belarus, Office of the President, "Mutual Relations of Belarus and Venezuela Must Develop into a Long-Term and Many-Sided Partnership," July 25, 2006, www.president.gov.by.

12. Ibid.

13. Ioffe, "Unfinished Nation-Building in Belarus and the 2006 Presidential Election," 37.

14. For the point/counterpoint, see Ioffe, "Unfinished Nation-Building in Belarus and the 2006 Presidential Election"; and Marples, "Elections and Nation-Building in Belarus: A Comment on Ioffe." Marples points out that a large part of Lukashenka's "success" owes to conveying, processing, or reselling Russian natural gas and crude oil, the profits of which have declined sharply as the Russians have raised their prices and forced Belarus to cede part-ownership of its pipeline network.

15. United States, Central Intelligence Agency, *World Factbook: Belarus*, May 15, 2008, 37–38.

16. CONAPRI [Consejo Nacional de Promoción de Inversiones], "Venezuela, Belarus Install Bilateral Committee" (Caracas, September 7, 2006).

17. BELTA [Belarusian Telegraph Agency], "Belorusian-Venezuelan Joint Venture to Extract Around 0.9 Tonnes of Oil in 2008" (Minsk, May 20, 2008).

18. RIA Novosti, "Belarus, Venezuela May Produce 1 Mil Mt of Oil in 2008," December 5, 2007, at www.rian.ru.

19. Belarus Office of the President, "Belarus Needs to Expand Its Presence in Latin America," December 4, 2007, www.president.gov.by.

20. "Baby Formula Drought Irks Belarus Mothers—Chávez to Blame?" Third Way Internet Community, November 15, 2007, at www.monstersandcritics.com/news/europe/.

21. Simon Araloff, "The Rise of Contemporary Belarus Military Industry," Axis Information and Analysis, July 7, 2005, at www.axisglobe.com.

22. Calvin Sims, "Peru's Cut-Rate Fighter Jets Were Too Good to Be True," *New York Times*, May 31, 1997.

23. "Belarus to Provide Venezuela Night Vision Devices," EUX-TV (from Belapan), March 23, 2007.

24. "Asesores militares viajarán este año a Venezuela para montar su defensa aérea," Terra Web site, April 4, 2008, at www.terra.es.

25. Maite Rico, "Los papeles de las FARC acusan a Chávez," *El País*, May 10, 2008.

# 7

# Responses to Venezuelan Petro-Politics in the Greater Caribbean

ANTHONY P. MAINGOT

"Bolivarian bravado" is the term used by Benedict Mander of the *Financial Times* to describe the rhetoric coming from the regime in Venezuela in the second half of 2008.[1] In an analysis supported by other serious journalistic accounts,[2] Mander pointed to several underlying realities of the situation at the end of 2008. First, he noted that while Venezuela was second only to Saudi Arabia in terms of oil reserves, these are, of course, in the ground, offshore, or in the difficult terrain of the Orinoco Belt, and will require foreign, that is, U.S., technology to fully develop. President Hugo Chávez's overwrought anti-American rhetoric appears to ignore this problem, Mander pointed out, as well as the fact that the United States is one of the few countries that pays for oil in cash up front. Were it not for this, Venezuela could hardly afford its enormous domestic and international outlays. The country's budget was planned on expectations of $60 per barrel of oil, and the November 2008 price was closer to $45. In addition, while Venezuela claimed to be producing 3.24 million barrels a day, according to OPEC the real figure was closer to 2.33 million—a not unusual discrepancy, since calculating the amount of oil any country produces is, in the words of the *Financial Times*, a "dark art."[3] Be that as it may, by mid-December Standard and Poor's had downgraded Venezuela's foreign debt three rankings below investment grade.[4]

This is the context in which Venezuela's assistance to seventeen countries in the Greater Caribbean should be analyzed. The major and most expensive part of this assistance is Petrocaribe, an influential instrument of Venezuelan foreign policy in the region. Petrocaribe was initiated under more favorable financial circumstances, but things have changed over the three distinct phases of its existence. The first phase began in July 2004 with the first meeting of energy ministers of the Caribbean in Caracas; the second phase in August 2004 with the second Petrocaribe meeting of energy ministers in Montego Bay, Jamaica; and the third in June 2005 at the energy meeting for the creation of Petrocaribe in Puerto La Cruz, Venezuela.

PetroCaribe is more than an energy-based agreement. It is intended to help integrate the Greater Caribbean region under a comprehensive program geared toward the "transformation" of these societies. The agreement recognizes the need for "special and differentiated" treatment of less-developed countries. It guarantees absolute respect for the principles of sovereignty, non-interference, and equality of states, with all terms and conditions determined through bilateral negotiations.

It all seemed too attractive for any country in the Greater Caribbean to turn down, especially considering the structural conditions of virtually all of these small states. All of the nations in the region, with the exception of Venezuela and Trinidad and Tobago, are energy dependent. As the foreign ministers who met at the Organization of American States (OAS) General Assembly meeting in Panama in 2007 acknowledged, "Energy is an essential resource for the sustainable development of peoples. . . . Access to energy is of paramount importance." Already in 1980 Mexico and Venezuela had inaugurated the San José Accord, based on an oil facility geared toward reducing the financial burden of energy costs for eleven Central American and Caribbean countries when the price of oil exceeded $15 per barrel. That accord is today in frank decline for two reasons: first, Mexico's declining production; and second, Chávez's promotion of his own energy aid program, Petrocaribe, which is far more generous.

Although the Caribbean received trade preferences under the Caribbean Basin Recovery Act (CBERA), it was really the Caribbean Basin Trade Partnership (CBPTA) which provided opportunities for products not covered by CBERA. CBPTA is due to expire in September 2010. Even if well-intentioned members of the U.S. Congress, such as Charles Rangel of the powerful Ways and Means Committee, seek ways to promote trade with the Caribbean, some private-sector interests think otherwise, and have even more influence in Washington. An example is the recent World Trade Organization (WTO) ruling that the European Union preference program for Eastern Caribbean bananas was not in compliance with international trade rules and had "harmed American trade rights." While the United States does not grow or export any bananas, three powerful U.S. companies do: Chiquita Brands International, Del Monte Foods, and the Dole Food Company. Even though the EU appears to want to assist its former colonies, and both sides have made great strides regarding tariffs, taxes, and duties, the Caribbean is ill-prepared to comply with stringent European phyto-sanitary conditions.

This tour d'horizon of the region reveals nation-states struggling to survive and develop in a world of ever-increasing energy costs, competition from lower-wage areas, and reduced geopolitical leverage since the end of the Cold War. Given these circumstances, it is unreasonable to expect them to reject any meaningful assistance, whether from the EU, the U.S., Venezuela, or Cuba. The circumstances provide opportunities to those powers that have the capacity to assist, with defined geopolitical designs in mind. Petrocaribe is part of that design in a way that the San José Accord was not.

## Petrocaribe: A Brief Description of Terms

Consider the terms of Venezuela's PetroCaribe Energy Cooperation Agreement:

1. Oil is not sold at concessionary prices. OPEC rules prohibit sales below world market prices by member states.
2. Instead of paying the full amount up front, only a part is remitted immediately in cash. The rest of the payment is converted into a loan, used for development purposes, to be paid off in twenty-five years at 1 percent or 2 percent interest.
3. No conditions are attached to the loan.
4. Arrangements can be made to repay portions of the loan in the form of goods and services.
5. All transactions are state-to-state. No private intermediaries are allowed.
6. Transportation and docking facilities as well as upgrades to refining and storage capabilities are included.

Is it any surprise that the majority of Caribbean nations signed on to the agreement? The only holdouts are Trinidad and Tobago and Barbados. Both cases call for some analysis.

## Trinidad and Tobago: The Non-Participant

Trinidad and Tobago's reservations concerning Venezuelan initiatives are not new. In fact, the current attitude toward Petrocaribe recalls the heated controversy that broke out between the two countries in the mid-1970s.[5] At the time, Trinidad and Tobago's prime minister Eric Williams objected to what he regarded as Venezuela's improper designs on the Caribbean region in general, and Trinidad and Tobago in particular. Two speeches he gave in 1975 illustrate Williams's concern.

In the first speech (May 1975), Williams attacked the notion that Venezuela was a Caribbean country ("I expect next to hear that Tierra del Fuego is") and pointed to "Venezuela's relations, territorial ambitions in respect of our area." The second speech was delivered to his party's convention on June 15, just two days before the prime minister was scheduled to leave on a trip to Cuba (continuing on to the USSR, Romania, and the United States, where he met with Henry Kissinger). In what amounts to one of the most scathing attacks by one Caribbean country on another during peacetime, Williams warned of Venezuela's "penetration" of the Caribbean, berated that country for its "belated recognition of its Caribbean identity," and chastised his CARICOM partners for falling for the new Venezuelan definition of the Caribbean Basin, and for leading a "Caribbean pilgrimage to Caracas."

The sources of Williams's irritation with Venezuela were many, and some were certainly legitimate. First, contrary to the provisions of the CARICOM Charter, which calls for multilateral trade with non-members, Venezuela was—as it is now—encouraging bilateral deals. This was especially the case for bauxite and oil, commodities that Williams himself wanted to dominate. Second, the two countries had differences regarding the laws of the sea, Venezuelan claims to Guyanese territory and certain islets in the Caribbean (among them Aves Rock, off Dominica), and Venezuelan loans, tourism initiatives, and scholarships. Williams expressed the fear that Caribbean and Latin American primary commodities were "jumping from the European and American frying pan into the South American fire," and that the net result would be the recognition of Venezuela as a "new 'financial centre' of the world."

The issues may vary today, and certainly the language is more diplomatic, but there is no doubt that Petrocaribe represents, if not a threat, then at least serious competition to the oil- and gas-based economy of Trinidad and To-bago. If we consider that Trinidad and Tobago's PetroTrin refinery makes 56 percent of its sales in the Caribbean, then the commercial threat is evident. When Venezuela supplied St. Vincent with 7,200 10-kg liquefied propane gas tanks under Petrocaribe, it was at the expense of Trinidad and Tobago's gas processors.

In addition to its commercial interests, Trinidad and Tobago had pur-sued its own "petro-diplomacy" in the region. The arrangements were lon-ger term than those of Petrocaribe: a rebate on oil purchases deposited in a collective fund meant for developing the economies of CARICOM countries.

Finally, Trinidad and Tobago perceives Petrocaribe, in conjunction with ALBA, as representing a challenge to its aspirations to be a major player in the Caribbean. Three major foreign policy initiatives were meant to facilitate this role:

- A cable carrying electricity to Grenada
  Closer ties with that island are a longstanding plan, and Trinidad is home to a large population of Grenadians that has considerable political clout.
- An Eastern Caribbean gas pipeline
  This plan has involved Trinidad and Tobago in high-level talks with France, since it includes the French islands of Martinique and Gua-deloupe.
- A major liquefied natural gas project in Jamaica

To carry out these monumental projects, Trinidad and Tobago will need foreign private investments (specifically excluded by Petrocaribe terms), the consent of the other islands (which are hardly in a position to wait that long), and an agreement with Venezuela on sharing gas fields in the Gulf of Paria. Success may seem quite distant and improbable at this point, but this has not stopped that ambitious island from thinking big and resenting potential competitors.

None of this, however, should be taken to mean that Trinidad and Tobago has strained relations with Venezuela. At a 2007 meeting in Caracas, Trini-dad and Tobago prime minister Patrick Manning and President Chávez an-nounced that the two countries had joint commissions studying the issue of oil and gas deposits in the narrow Gulf of Paria.[6] This continues a tradition of deft foreign policy initiatives, more akin to Brazil's *pragmatismo responsavel* than to the U.S. propensity toward a "with us or against us" approach.

It should be remembered that in 1972 Eric Williams led the move by the four independent CARIFTA states (Trinidad and Tobago, Jamaica, Barbados, and Guyana) to establish relations with Cuba. By 1993, all CARICOM states had diplomatic relations with Cuba, and a CARICOM-Cuba Joint Commission coordinates activities in a range of areas. These ties with Cuba serve Trinidad and Tobago well when relations with Venezuela become difficult. Trinidad and Tobago has learned well the art of diplomatic triangulation.

## Barbados Responds Coolly

Trinidad and Tobago has found additional satisfaction in the fact that Barbados, generally considered the best-governed of the CARICOM islands, has also refused to join Petrocaribe. There are several reasons for this. First, Barbados has always had a close relationship with Trinidad and Tobago and does not wish to affect this friendship in any way. Second, Barbados is now one-third self-sufficient in oil and wishes to attract more foreign investments for exploration. In early May 2008, Barbados opened trading on the rights to explore twenty-eight offshore "blocks," with bidding expected by some thirty companies. When Venezuela claimed that two of these blocks were in its waters, a repeat of the Trinidad and Tobago–Venezuela disputes of the 1970s appeared to be in the making.[7] Finally, the economy of Barbados is sustained by tourism and the financial services sector, hardly items propitious for a barter arrangement. With little in terms of native products to provide as exchange, the island was apprehensive about incurring heavy debts from Petrocaribe loans.

The Barbados case provides insights into the way a country arrives at foreign policy decisions based on a wide range of political and economic systems and customs. Steady resoluteness is a Barbados cultural trait, and it is demonstrated quite clearly in the island's response to Petrocaribe.[8]

## Jamaica: The Pragmatic Borrower

Jamaica is heavily energy-deficient, importing 90 percent or more of its energy needs. This fact alone goes a long way toward explaining the compulsion the island felt to secure the steady flow of oil promised by Petrocaribe. In 2004, the year before it signed the PetroCaribe agreement, Jamaica spent more than 60 percent of its export earnings on petroleum products. The price of a barrel of crude in 2004 was $34, double what it had been in 2001. Using 1987 as a base year, Jamaica's GDP had grown by 20 percent, but energy consumption increased by 112 percent.[9] Part of the Petrocaribe agreement was to upgrade the island's PetroJam refinery by about 42 percent, to 50,000 barrels per day. The prime minister at the time, P. J. Patterson, summed up the global context facing his island and why the 23,500 barrels of oil a day imported from Venezuela were so welcome: "A new corridor has been created for us in the Caribbean to supply to Venezuela certain goods and services that may be affected by emerging trade policy, including decisions of the WTO which are inimical to member states."[10]

Trinidad and Tobago was not at all happy to see Jamaica cozying up to Venezuela. In early 2007 it announced that it was not going ahead with the previous arrangement to supply liquefied natural gas to Jamaica and advised it to approach Venezuela instead. On March 13, 2007, Jamaica signed a memorandum of understanding with Venezuela for 150 million cubic feet of liquefied natural gas per day. In turn, the Jamaican Manufacturers Association criticized the Trinidad and Tobago government for reneging on its agreement and for making it increasingly difficult to export Jamaican goods to that island, despite the fact that Trinidad and Tobago had a US$ 500 million favorable balance of trade with Jamaica.[11]

While Jamaica certainly benefited from the Petrocaribe deal and especially from the Petrocaribe Development Fund, in no way did this change the dynamics and orientation of Jamaican politics. In 2007 the Jamaican people voted out the People's National Party (PNP) government that had signed the agreement, replacing it with the more conservative Jamaica Labour Party (JLP). Petrocaribe was never an issue during the campaign. The new prime minister, Bruce Golding, has kept the agreement going as a purely "business" arrangement, proof that whatever geopolitical and ideological intentions Chávez might have had have not materialized. Jamaica learned a painful lesson during the 1970s when Michael Manley made a sharp turn to the left, only to see the economy, and his political base, collapse. Jamaica's politics, and its foreign policies, are today geared toward a pragmatic search for solutions to the island's many domestic problems.

## Dominica: An Eager Participant

In February 2007, two weeks before the CARICOM heads of government were to meet in Washington, President Chávez visited Dominica and St. Vincent. In St. Vincent he announced plans for a new US$200 million airport to be built with Venezuelan funds and Cuban labor. Typically, he took advantage of the occasion to declare, "Down with U.S. imperialism! Long live the people of the world!" An AP journalist present noted, "The crowd did not respond with applause to the Venezuelan leader's vitriolic statements."[12]

The author does not know what Chávez said in Dominica, but given Prime Minister Roosevelt Skerrit's decision to sign on to Petrocaribe, become the only CARICOM member to join ALBA, and to strengthen ties with Cuba, some observers warn of an American "backlash" à la Grenada, 1983. They speak of nations "expectantly entering into Venezuela's geopolitical orbit."[13] These alarms are simplistic and alarmist to an extreme.

Dominica is one of the poorest islands in the Eastern Caribbean. This poverty has been aggravated by the decline of the banana industry. In 2006, Dominica's total tax revenue was EC$194 million and its expenditures EC$270 million, deepening a years-long budgetary shortfall. What economic opportunities are available for this 290-square-mile island with a population of 65,000 to "balance its books," as the thrifty and conservative folk on the island would say?[14] They are not optimistic about the prospect of U.S. aid. Dominicans know well that it was the United States, protecting its banana companies based in Ecuador and Colombia, which brought the WTO suit against the United Kingdom's preferential prices for Eastern Caribbean bananas. This loss of market for its main product, plus the losses wrought by hurricanes (damages from Hurricane David alone were 20 percent of GDP) have left the island with few options. Unlike other islands, where an expanding tourist industry has compensated for the decline in agricultural exports, Dominica's volcanic, mountainous terrain (61 percent of the land is mountainous and forested), while ideal for ecotourism, is not conducive to mass tourism projects. There is a distinct lack of white sandy beaches, and the airport is not easily accessible. None of this, of course, stops Venezuela from promoting Bolivarian "social tourism" in Dominica.[15] This politically driven exercise in once-a-week subsidized group tours is hardly what the island's economy needs, or what tourist authorities wish to promote. With so few alternatives, joining ALBA makes sense to even the most conservative Dominican sectors, including the island's Chamber of Commerce.

The Dominica-Venezuela link is sustained by eleven cooperative programs. Many of these are noncontroversial, including the improvement of housing damaged by Hurricane David, the paving of mountain roads, upgrading the agricultural sector, and increasing the number of university scholarships. Other programs, however, have met with strong opposition. These include the plan to build a refinery on the northeastern coast of Dominica. Opponents argued that the deal clashed with the national plan to promote ecotourism and that it was signed without any public consultation, putting at risk Dominica's relations with other countries, especially the U.S. and Trinidad and Tobago.[16] Widespread public protests forced the Dominican government to put the project on hold. Another source of controversy was Venezuela's donation of US$4.5 million to build houses and schools for Dominica's Carib Indians. There can be no doubt that the Caribs need help, but opponents feared that Venezuelan money would dramatically alter the communal basis of Carib society. In the end, the aid was accepted, considering that the only alternative proposed by the Dominican government to address the high levels of alcoholism and drug addiction on the reservation was "Prayers and Bible Study . . . creation of Godliness and brotherly love."[17]

Dominica has a vibrant two-party system and a pragmatic foreign policy that takes into account its debts to Trinidad and Tobago, Barbados, and Martinique, home to a large emigrant community. It has put on hold its dispute with Venezuela over neighboring Aves Rock and has shown itself willing to make some concessions to the Bolivarian initiative. Ultimately, however, the island can count on built-in democratic safeguards and a political system that insists on transparency and accountability.

### Nicaragua: New Life for the Sandinista Revolution

Few events illustrate more dramatically the gap between expectations and actual benefits from Petrocaribe than the Nicaragua transport strike of May 2008. This stoutly Sandinista union was demanding subsidized fuel to meet escalating costs. The union eventually succeeded in its demands, but without any direct help from Petrocaribe; instead, the funds came from donations from "other" ALBA members, which could mean only one thing: Venezuelan funds authorized by President Chávez himself.[18]

After Cuba, Nicaragua has been the most favored recipient of Venezuelan largesse. The favors have come in the form of debt forgiveness, donations of tractors, an aluminum plant, an oil refinery, electrical generators, construction of a Pacific-Atlantic highway, and a wide range of medical services, including Operación Milagro (Operation Miracle), a free surgery program.

Nicaragua's participation in Petrocaribe provides it with 10 million barrels of oil a year, enough to cover its energy needs. Nicaragua pays for 50 percent of the oil up front, with the rest to be paid off in twenty-three years at 2 percent interest. In the meantime, the money is meant to be used as a loan toward infrastructure and development. None of it is registered as public funds, since it is handled by an "autonomous" agency (Albanisa) and the state-owned petroleum company, Petronic. Both Albanisa and Petronic are administered by Sandinista partisans.

This Venezuelan slush fund greases the Sandinista patronage system, which is administered through a network of Citizens Power Councils controlled by First Lady Rosario Murillo and widely perceived as preliminary steps in the creation of "Bolivarian direct democracy."[19] A pervasive sense of *personalismo* is everywhere evident, deepened by the fact that the Sandinista party secretariat is located within the Ortegas' private residential compound. Beyond its increasingly authoritarian domestic moves, the Sandinista government has renewed an aggressive foreign policy emphasizing rhetorical support for other embattled ALBA members, especially Bolivia. Tensions continue with Colombia over the San Andrés and Providencia islands despite the International Court of Justice's decision in favor of the South Americans.

In response to the current situation, several European countries have suspended foreign aid to the Nicaraguan government. After Germany decided to cut off its assistance, the German newspaper *Suddeutsche Zeitung* cited "deficiencies in the division/separation of powers, the subordination of the justice system to politics, the lack of transparency in the handling of public funds, corruption, and disdain for human rights."[20] None of this, of course, was a creation of Chávez or Bolivarian ideology. A rereading of original Sandinista documents confirms the profoundly authoritarian strain of Sandinista ideology well before the emergence of *bolivarianismo.*[21] For an understanding of "twenty-first-century socialism," an April 13, 2007, editorial in *La Prensa*, the nation's most circulated newspaper, noted, "Just look at the twentieth-century Sandinista socialism that misgoverned the country between 1979 and 1990."

## Conclusion

As the cases above make clear, every country reacts to Venezuela and its Bolivarian revolution in its own way. Existing political cultures—even in relatively new states—are not easily modified, even in the face of the most attractive incentives. Trinidad and Tobago, Jamaica, and Dominica have chosen the parliamentary system and, judging from recent elections, it works well in alternating parties in power. Since January 2006, six governments have changed hands: in the Bahamas (2007), Barbados (2008), Haiti (2006), Jamaica (2007), and St. Lucia (2006). All of the political leaders have been elected through the democratic process.

All of these countries have normal relations with Cuba and have been courted, individually and collectively, by President Chávez. Past experience and research lead one to deduce three major reasons why these small nations will never bite a hand that feeds them, but will hardly change their system to suit it, either. First, while virtually all West Indian parties had their origins in the trade union movement, and many adhered to Fabian-type socialism, dramatic events in the 1970s and 1980s changed the perception of the word "socialism." It still has an ominous ring in the Caribbean. The tragedy of Grenada, the assassinations in Suriname, the failure of Michael Manley to show that a "humanistic" socialism was possible, all soured the region on that word.[22] Even Cuba, which engenders much sympathy and friendship throughout the region, is seen more as a case of nationalistic determination in the face of U.S. aggression and less as a shining example of economic development. Second, despite the evident language barriers, contemporary Caribbean politicians have passed the stage of the occasional poorly educated (albeit not unsophisticated) labor leadership, and tend to elect leaders with advanced levels of education. They find President Chávez's "twenty-first-century socialism" confusing and ultimately unconvincing.

This ideology has at its core the idea of creating an alternative (Alternativa Bolivariana para las Américas, or ALBA) to U.S. imperialism, capitalism (in its neoliberal, free trade, and privatization dimensions), and traditional oligarchies (often defined in racial terms). Its basic tenets were formulated by the German-Mexican Marxist Heinz Dieterich, whose book *Socialism of the*

*XXI Century* Chávez adopted as his ideological guide. After a period of ideological euphoria, Dieterich himself began accusing Chávez of a series of missteps that endangered not only his own revolution but those in Cuba and Bolivia. According to Dieterich, the *caudillo*-style leadership of the ALBA countries contradicts the fundamental tenet of his model: that socialism has to grow from the bottom up, never from the top down, and never so centralized.[23] This is exactly what one finds in the "Bolivarian" regimes of Chávez, Correa, Morales, and Ortega.

The contradictions, both philosophical and practical, of Chávez's ideology have been duly noted by Venezuelans who know something about socialism and international affairs. Ex-guerrilla fighter and present leader of the Movimiento al Socialismo (MAS) in Venezuela, Teodoro Petkoff, claims that few if any of Venezuela's socialist intellectuals understand, much less support Chávez, his ideology, and his regime's policies.[24]

The most attractive and successful aspect of the Bolivarian initiatives is medical assistance. Operación Milagro has to be ranked as one of the most generous and appreciated uses of "soft power" in the Greater Caribbean, and perhaps all of Latin America. The problem for Chávez lies in the fact that even though he funds the initiative, those in the front lines are Cuban, and they and their country get the credit for their good work.

Beyond oil and medical attention, however, Venezuela's "soft power" program of material and ideological inducements cannot compete with the attractions and "pull" of U.S. society. Although the Greater Caribbean has long disliked and distrusted U.S. power (so often abused in the past), it has no quarrel with a series of other areas that are the real sources of U.S. soft power: popular culture, technological modernity, admiration for its democratic politics, and the fact that the United States is the country with the largest number of immigrants proportionally and the easiest in which to acculturate and assimilate, as well as to secure citizenship.

There is nothing to guarantee, however, that the attraction of U.S. culture will continue. As Joseph Nye has said, noting the recent U.S. penchant for using "hard power": "We have been less successful in the areas of soft power, where our public diplomacy has been woefully inadequate and our neglect of allies and institutions has created a sense of illegitimacy that has squandered our attractiveness."[25]

In the final analysis, the sophistication of Caribbean democrats is the main barrier to Chávez's lack of intellectual sophistication and penchant for braggadocio. This is the main lesson to be learned from Venezuela's efforts in the region: countries will seldom turn down economic assistance when they need it, but how they respond politically to that assistance will depend on their own political culture and realities.

## Notes

1. Benedict Mander, "Bolivarian Bravado," *Financial Times*, November 28, 2008.

2. *New York Times*, December 17, 2008.

3. *Financial Times*, November 28, 2008.

4. *Alexander's Gas and Oil Connection* 6, no. 16 (August 28, 2001).

5. This section is taken from Anthony P. Maingot, *The United States and the Caribbean: Challenges of an Asymmetrical Relationship* (London: Macmillan; Boulder, Colo.: Westview Press, 1994), 131–32.

6. "Patrick Manning visita Caracas," *El Universal*, March 20, 2007, http://www.politica.eluniversal.com/2007/03/20/pol.

7. *Daily Nation* (Barbados), June 18, 2008. A heavy debt burden from Petrocaribe oil is presently a problem in the Dominican Republic. The big issue at this writing is whether the government's buyout of Shell's 50 percent of the island's only refinery (Refidomsa) to increase the flow of Venezuelan oil will pay off. Doubts about the state's managerial skills and its capacity to repay Venezuela are the main issues being debated. A comparison between the ways Barbados and the Dominican Republic arrived at their respective decisions would be instructive.

8. Anthony P. Maingot, "The Paradoxical Origins of Barbados' Civic Culture," lecture delivered at the Culture Change Institute, Fletcher School of Law and Diplomacy, October 24, 2008.

9. Sixty percent of Jamaica's petroleum imports went into electricity generation, mining (bauxite, alumina), and manufacturing.

10. The Most Honorable P. J. Patterson, statement to Parliament, July 13, 2005, at www.jis.gov.jm/special_sections/CARICOM.

11. *Jamaica Gleaner*, March 20, 2007.

12. Duggie Joseph, "Chávez in St. Vincent Calls for Anti-Imperialist Unity," Associated Press, February 17, 2007.

13. Nikolas Kozloff, "Dominica: The Caribbean's Net 'Terror Island?'" Council on Hemispheric Affairs Web site, February 26, 2008, at www.coha.org.

14. On the conservative nature of Caribbean culture, see Anthony P. Maingot, "The Caribbean: The Structure of Modern-Conservative Societies," in *Latin America: Its Problems and Its Promise,* ed. Jan Knippers Black (Boulder, Colo.: Westview Press, 1998), 436–52.

15. *Agencia Bolivariana de Noticias*, March 14, 2008, www.abn.info.ve.

16. "Freedom Party Slams Skerrit's ALBA Agreement," *The Sun* (Dominica), February 25, 2008.

17. Ministry of Community Development, *Heritage of the Kalinago People* (Roseau, Dominica, 2007).

18. *El Nuevo Herald*, May 19, 2008.

19. James C. McKinley Jr., "Nicaragua Councils Stir Fear of Dictatorship," *New York Times*, May 4, 2008.

20. See the essay by the Costa Rican columnist Eduardo Ulibarri, "Ortega y los cangrejos," *El Nuevo Herald*, May 7, 2008.

21. Robert S. Leiken and Barry Rubin, eds., *The Central American Crisis Reader* (New York: Summit Books, 1987), 208–318.

22. See Anthony P. Maingot, "The Difficult Road to Socialism in the English-Speaking Caribbean," in *Capitalism and the State in US-Latin American Relations,* ed. Richard R. Fagen (Stanford, Calif.: Stanford University Press, 1979), 254–89; Anthony P. Maingot, "Political Processes in the Caribbean, 1970s to 2000," in *UNESCO General History of the Caribbean,* ed. Bridget Brereton (London: Macmillan, 2004), 5:312–45.

23. "Entrevista a Heinz Dieterich," *Diario La Tercera*, March 29, 2007; Pedro Páramo, "Chávez y el padre del socialismo del siglo XXI," *Contrapunto* (January–March 2008): 23–25.

24. Teodoro Petkoff, "Una postura inmoral," El Tiempo.com, February 23, 2008, at *www.eltiempo.com.*

25. Joseph S. Nye Jr., *Soft Power* (New York: Public Affairs, 2004), 147.

**8**

# ALBA, Petrocaribe, and Caricom

Issues in a New Dynamic

NORMAN GIRVAN

The growth of relations between several Caricom states and the Venezuelan-promoted ALBA and Petrocaribe initiatives is one of the most significant recent developments in regional affairs. An immediate issue that has arisen is whether membership in ALBA conflicts with the obligations of Caricom membership. Larger issues of a strategic nature include the need for diversification of economic relations in light of global economic restructuring; pursuit of opportunities for South-South cooperation that are more advantageous to the region than standard North-South arrangements; and the scope for a coordinated Caricom external trade policy.

Although ideology and hemispheric geopolitics come into play in any discussion of ALBA and Petrocaribe, it is important to frame the issues within a regional optic rather than from the point of view of Washington. ALBA, in addition to having its own special characteristics, can be seen as one manifestation of a process of reconfiguration within the world political economy, a process marked by a relative decline in U.S. power and the emergence of new geo-economic poles of influence. The rise of Asia, and in particular China and India, is among the most significant of these changes, as is the emergence of other regional powers in the global South, such as South Africa, Brazil, and Venezuela. One notable consequence is the waning ability of the United States to control the course of events in Latin America and the Caribbean. As a recent report published by the Washington-based Council on Foreign Relations noted, "The era of US hegemony [in the region] is over."[1]

The signs of this shift are everywhere. The Free Trade Area of the Americas (FTAA) process has stalled over Brazilian opposition to Washington's terms for the negotiations; governments opposed to the neoliberal "Washington Consensus" have come to power in several countries; the Cuban Revolution is about to celebrate its 50th anniversary despite Washington's obsession with regime change in that country; the Bolivarian revolution in Venezuela continues apace; and the traditional Washington-dominated sources of development cooperation are being overshadowed by Southern-controlled institutions centered on Venezuela and Brazil. At the continental level, a South American Union (Unasur) is being constructed under Brazilian leadership. These developments form an important backdrop to a consideration of the role and significance of ALBA and of Caricom's relationship with the grouping.

This chapter examines the nature of ALBA's mission and program, focusing on the kind of cooperation arrangements that are likely to be of particular interest to Caricom countries. It reviews the scope and magnitude of financial cooperation, the existence of non-reciprocity, the scope of social cooperation, the role of Petrocaribe, and the recent incorporation of food security into the ALBA cooperation program. The discussion includes the content of ALBA agreements from the point of view of the treaty obligations of Caricom members and reaches the conclusion that there is no inherent incompatibility between them. Also considered are the potential economic and political vulnerabilities of participation in ALBA and Petrocaribe, with suggestions as to how these might be mitigated. Finally, this chapter emphasizes the value of coordinated Caricom policies on ALBA and other external economic relations, and discusses the difficulties of agreement in a community with divergent interests among its members.

## The ALBA Mission

The Bolivarian Alternative for the Americas, or ALBA, presents itself as an alternative to the U.S.-sponsored neoliberal model of economic integration based on trade and investment liberalization.[2] ALBA claims to put the basic needs of the population and the reduction of poverty above private profits and the rights of private investors. The guiding principles of ALBA integration are said to be solidarity, complementarity, compensatory financing for the treatment of asymmetries, and differentiated treatment of

countries according to their circumstances. In practice, ALBA's cooperation has consisted mostly of concessional financing for the relief of energy
import bills, state-owned industries, and physical and social infrastructure; support for health and education projects that directly benefit the
poor; and non-reciprocal trading arrangements.

ALBA has grown significantly since it was launched by Venezuela and Cuba
in December 2004. Bolivia, Nicaragua, and Dominica have acceded, and St.
Vincent and the Grenadines and Antigua and Barbuda have signed statements

### ALBA Principles and "Rules"

1. Trade and investment are to be instruments of fair and sustainable development with effective state participation.

2. Special and differentiated treatment for participating countries
according to their level of development and size.

3. Economic complementarity and cooperation between countries to preserve efficient and productive specialization and balanced economic development.

4. Cooperation and solidarity for a continental fight against illit-
eracy and to provide free health care and scholarships.

5. Creation of a Social Emergency Fund.

6. Integrated development of communications and transport.

7. Environmental protection.

8. Energy integration.

9. Promotion of intra-Latin American investment through a Latin
American Investment Bank, a Southern Development Bank, and a
Latin American Society of Reciprocal Guarantees.

10. Defense of cultural identity; creation of Telesur.

11. Intellectual property rights to protect regional patrimony
without posing an obstacle to cooperation.

12. Harmonization of positions in multilateral forums, including
democratization of international organizations, particularly the
United Nations system.

Acuerdo entre el presidente de la República Bolivariana de Venezuela y el presidente
del Consejo de Estado de Cuba, para la creación de la Alternativa Bolivariana para las
Américas, December 14, 2004, http://www.alternativabolivariana.org/modules.php?name
=Content&pa=showpage&pid=2060

of support. The Sixth ALBA Summit, in January 2008, was attended by the leaders of twelve countries,[3] including nine from the Greater Caribbean and five Caricom member states. According to the Venezuelan Ministry of Integration and Foreign Trade, the main initiatives of 'Caribbean ALBA' are:

1. The fight against poverty and social exclusion
2. A joint plan for food security
3. Power and mining development for creating joint production chains
4. A portfolio of integrated investments
5. Academic and cultural exchange
6. Tourism cooperation in human resource development, air transport, and other areas
7. Environmental conservation
8. Caribbean regional trade
9. Prevention and management of natural disasters

### ALBA Agreements

December 14, 2004: Joint Statement by the Presidents of the Bolivarian Republic of Venezuela and Cuba for the Creation of ALBA

December 14, 2004: Agreement between the Presidents of the Bolivarian Republic of Venezuela and Cuba for the Application of ALBA

December 14, 2004: Contribution and Subscription of the Republic of Bolivia to the Joint Statement

April 29, 2006: Agreement for the Application of ALBA and the People's Trade Treaty (TCP) between Bolivia, Cuba, and Venezuela

May 17, 2006: Agreement in the Framework of ALBA and the TCP for the Instrumentation of Special Financing Funds

January 11, 2007: Accession of Nicaragua to the Joint Statement and Statement of Contribution and Accession of Bolivia

February 17, 2007: Joint Statement of the President of Venezuela and the Prime Ministers of Antigua and Barbuda, Dominica, and St. Vincent and the Grenadines

January 26, 2008: Document of Accession to ALBA signed by the Prime Minister of Dominica

The original Spanish text of many of these agreements is available at http://www.alternativabolivariana.org.

## ALBA Financial Cooperation

Financial cooperation is an important element of ALBA. Venezuelan made a commitment of $100 million to Bolivia when the latter joined the agreement, along with an additional $30 million for infrastructure projects. Two major new developments are the ALBA Caribe Fund and the ALBA Bank. The ALBA Caribe Fund was created within the Petrocaribe framework specifically for ALBA participating countries. Reportedly, 25 percent of the bill for imports of crude oil from Venezuela is credited to this fund, whose purpose is to fight poverty by financing social and economic programs. Information on the amount accruing to this fund and the level of disbursements is not available, but given the steep increases in oil prices, it is likely to grow rapidly.[4]

Bolivia, Cuba, Nicaragua, and Venezuela agreed to establish the ALBA Bank (BALBA) in June of 2007, and the bank was formally launched in 2008. Its objectives are to support sustainable social and economic development, reduce poverty, and strengthen integration. BALBA started with subscribed capital of $1 billion but has authorized capital of twice that much.

## Non-Reciprocity in ALBA

Non-reciprocity and compensated trade (i.e., through direct product exchanges) are two ways in which the principles of fair trade and special and differentiated treatment are applied in ALBA. Furthermore, trade agreements are negotiated on a case-by-case basis, allowing for flexibility of commitment according to each country's circumstances. These principles are broadly applied; for instance, some non-reciprocal features of the Cuba-Venezuela agreements actually favor Venezuela, even though it may be considered the "more developed" member. Hence, Cuba agreed to grant duty-free access to Venezuelan imports and to remove non-tariff barriers, while in return Venezuela has agreed to eliminate only non-tariff barriers on Cuban imports. This imbalance is in recognition of the fact that "Venezuela is a member of international institutions that Cuba does not belong to, all of which must be taken into consideration when applying the principle of reciprocity in the commercial and financial arrangements that are made between the two countries."[5] Similarly, Bolivia has been granted duty-free access to Cuba and Venezuela and elimination of non-tariff barriers to its exports without incurring the same obligation in return.[6]

In payment arrangements, there is provision for payment-in-kind for oil imports from Venezuela ("compensated trade") and for reciprocal credit arrangements, both of which are contained in the Venezuela-Cuba agreement.[7]

Non-reciprocity in payment obligations also applies: Bolivia can pay for Cuban imports with Bolivian products, with the national currency of Bolivia, or with other mutually-agreed-upon currencies, but is not obliged to grant reciprocal terms to Cuba.[8] Venezuela has been granted the same facility in paying for Cuban imports, without reciprocity from Venezuela.[9] In the Caribbean, Dominica is allowed to pay for 40 percent of its Petrocaribe oil imports with exports of bananas.[10]

Non-reciprocity, however, is not always a good thing. The danger is that Caricom will come to be seen, and will come to see itself, as a "freeloader" in its external relations. It is in the interest of the community to identify ways in which it can assist other ALBA participating countries. English-language training and tourism-related training are two areas that come to mind. Caricom could also seek to establish its own technical assistance and volunteer programs for service in other developing countries, not restricted to ALBA. Non-reciprocity can, and should, be reciprocated.[11]

## Social Cooperation

Cooperation in health and education are major elements of ALBA, thanks in large part to Cuba's considerable human resource capabilities in these sectors. According to Cuban reports, some 30,000 doctors provide free services to the poor throughout Latin America and the Caribbean; 70,000 students are receiving training as health professionals; literacy programs have reached more than 2,000,000 people; and 600,000 patients have had their sight restored via Operation Miracle and free surgical operations.[12]

ALBA agreements provide 2,000 Venezuelan students with Cuban scholarships annually, and 5,000 Bolivians receive Cuban medical scholarships. Bolivia reportedly benefits from 600 Cuban medical specialists, and Venezuela has 15,000 Cuban medical professionals in its Barrio Adentro program.

More than 100 students from Dominica are reported to be attending Cuban medical and nursing schools, and approximately 75 Dominican students are in other Cuban educational institutions. Together, Cuba and Venezuela supply Dominica with about 2,000 scholarships in fields such as computer science, medicine, engineering, sports, physics, math, and agriculture.[13] Several hundred visually impaired Dominicans are said to have had their sight restored in either Cuba or Venezuela through Operation Miracle.

## Petrocaribe

Petrocaribe was initiated in June 2005 as an extension of the Caracas Energy Accord of 2001. As such, it predates ALBA and is available to sixteen countries in the Greater Caribbean, without requiring ALBA membership. Petrocaribe finances a portion of the value of imports of crude oil from Venezuela according to a sliding scale: above $30 per barrel, 25%; above $40, 30%; above $50, 40%; and above $100, 50%. The balance is payable over twenty-five years at 2%, falling to 1% at oil prices above $40 per barrel, with a two-year grace period for repayment.[14]

As the price of oil on world markets has grown, so has the value of Petrocaribe loans to importing countries. One consequence is that Petrocaribe has become the largest single source of concessional finance to the Caribbean region. From June 2005 to December 2007, Petrocaribe credits to importing countries amounted to $1.17 billion. They are expected to reach $4.5 billion by 2010.[15] This breaks down to $468 million per year in 2005–7, rising to $1.1 billion in 2008–10. By comparison, U.S. foreign assistance to the Caribbean region for fiscal years 2005–7 is estimated at $340 million per year ($149 million excluding Haiti).[16] Inter-American Development Bank disbursements to the sixteen Petrocaribe participating countries in fiscal year 2008 amounted to $100 million, less than one-quarter of Petrocaribe average lending for 2005–8.

For Caricom countries, the shift in the relative importance of sources of concessional finance is no less marked. Jamaica alone benefited from Petrocaribe lending to the tune of $471 million by the end of March 2008. U.S. assistance to Jamaica in fiscal years 2005–7 amounted to $58 million—that is, less than one-eighth as much. For the Eastern Caribbean and Suriname, the value of Petrocaribe credit is estimated at $180–$360 million per year,[17] compared to U.S. assistance of approximately $15 million in fiscal years 2005–7. IADB disbursements to all of Caricom in 2007 amounted to $43 million, a fraction of the Petrocaribe total. Petrocaribe also exceeds the EU's Regional Indicative Program for Cariforum countries by a wide margin: the tenth replenishment of the European Development Fund (EDF) is programmed at €165 million (approximately US$255 million), or $45 million per year.

Since the beginning of 2008, the skyrocketing price of oil on world markets has considerably enhanced the strategic role of Petrocaribe. At current import rates of 72 million barrels per year, each dollar rise adds $72 million per year to the oil bill of importing countries. At a base price of $30 per barrel, the recent world market price of $135 per barrel represents an additional $7.6 billion per year in debt for the sixteen importing countries. Petrocaribe credits could finance between 25 percent and 50 percent of this amount.

Access to Petrocaribe is not conditional on accession to ALBA; however, Petrocaribe shares many elements in common with ALBA, including compensation for asymmetries and the financing of poverty reduction and the state sector. For Venezuela, Petrocaribe and ALBA are expressions of a Bolivarian vision that includes other initiatives, such as Petroandina, Petrosur, Telesur, and the South Bank, which has authorized capital of $7 billion.[18]

ALBA countries plus Haiti also derive an extra benefit from Petrocaribe: 90 days credit for payment of half the value of oil shipments, with part payment in product exchange. For the balance, 25 percent of the import bill is extended as a direct credit to the government of the importing country, and 25 percent is paid into the ALBA Caribe Fund administered by PDVSA for social and economic projects within the importing country. The ALBA Caribe fund is a new institutional development within the ALBA Caribbean landscape and is destined to become a major player in regional financing.

## Food Security

An extraordinary ALBA summit, held on April 26–27, 2008, addressed the issue of rising food prices and food shortages in the region. The leaders of Bolivia, Cuba, Nicaragua, and Venezuela agreed to create the ALBA Network of Food Trade and the ALBA Food Security Fund, with an initial investment of $100 million. They also approved the creation of a commission comprising the agriculture and forestry ministers of ALBA member countries, with the objective of organizing joint productive projects and agro-industrial development in cereals, leguminous and oleaginous plants, meat, and milk.

In short, ALBA and Petrocaribe are significant developments in the hemispheric geo-economic and geopolitical landscape. Caricom countries cannot afford to ignore these developments; indeed, they are already strongly engaged on a bilateral basis. ALBA and Petrocaribe are major new sources of balance-of-payments relief in the face of rising oil prices, of financial assistance for government budgets and badly needed physical infrastructure, and of technical cooperation social services and human resource development. They have proven to be proactive in the face of new developments such as rising food prices. They are low-conditionality and involve a considerably lower degree of intrusion into domestic policies in scope and depth than the funding from traditional donors.[19] Let us now turn to the issues and risks involved in ALBA association.

## Do ALBA Obligations Conflict with Caricom Obligations?

The main issues to be considered here are the form of association with ALBA and the nature of the commitments assumed by acceding countries.

The inter-governmental modes of association employed in ALBA are those of Joint Statement, Agreement, Statement of Accession, Statement of Contribution and Subscription, and Statement of Support signed by heads of state and/or government. ALBA therefore does not take the form of an international or inter-governmental organization, treaty, or integration scheme in the normal sense. Adhering states do not agree to be legally bound by any statutes or obligations under international treaty law. "Principles" and "agreements" appear to be of a political nature; they are bilateral or trilateral documents to which specific heads of government subscribe. The term "membership," therefore, may be misleading as to its connotations in the case of ALBA (although it is used in both English- and Spanish-language documents). The terms "accession," "adherence," or "participation" may be more appropriate in conveying the nature of the association.

The question therefore turns on what specific obligations apply to acceding countries in general and to Dominica, as an acceding Caricom state, in particular. In the cases of Bolivia and Nicaragua, accession involved adherence to the Joint Statement by Cuba and Venezuela of December 14, 2004. This is a political declaration containing certain principles to which the governments subscribe. Some observers have suggested that the thrust of this statement is for the creation of an ALBA "economic bloc" (a term that is sometimes used in reference to ALBA) and that it will conflict with the implementation of the Caricom Single Market and Economy (CSME). A close reading, however, suggests that this is not the case.

## Petrocaribe Projects

Jamaica: Supply of 23,500 barrels per day. Agreements signed in education, science, technology, medicine, tourism, and for upgrading the Petrojam refinery.

Grenada: Agreements to supply 340,000 barrels per year in products: 55,000 gasoil, 85,000 gasoline, and 200,000 fuel oil.

Cuba: Inauguration of the Cienfuegos Refinery with a capacity of 70,000 barrels per day.

Belize: Mixed enterprise between PDVSA and Belize Petroleum.

Nicaragua: 80,000 gallons of Venezuelan diesel.

Dominica: 1,200 barrels of asphalt, warehouse; 1,000 barrels per day, hydrocarbons.

Antigua and Barbuda: Warehousing and distribution of fuel to the Eastern Caribbean.

New Projects:

Expansion of the Kingston refinery, Jamaica.

Construction of refineries in Nicaragua, Dominica, and Belize.

Completion of liquefied petroleum gas refinery in St. Vincent and the Grenadines.

Construction of fuel distribution plants in Dominica, St. Kitts and Nevis, St. Vincent and the Grenadines, Grenada, and El Salvador.

Electricity generation plants in Nicaragua, Haiti, St. Kitts and Nevis, and Antigua and Barbuda.

ALBA participants make no commitment to liberalize trade and investment, or adopt common economic policies, or to erect common economic barriers against the rest of the world. The normal features of an orthodox integration scheme are absent; hence the possibility of conflict with existing integration scheme obligations does not arise. Notably, both Venezuela and Bolivia are members of the Andean Group integration scheme; Venezuela is in the process of negotiating Mercosur membership; and Nicaragua participates in CAFTA, the Central American free trade agreement with the United States.

Most of the general principles of the Joint Statement appear to be unob-jectionable from the Caricom point of view. Two of them, however, might be considered questionable. No. 12 calls for "harmonization of positions within the multilateral sphere" and refers particularly to the fight to democratize the U.N. system. However, the ACS Convention also calls for harmoniza-tion of positions in international forums. All ALBA countries except Bolivia are also ACS members. The harmonization principle is always difficult to apply, because of differences among member states (even within Caricom there are difficulties) and because of the consensus rule in decision-making, which means that even one country can hold up harmonization. Further-more, ALBA is not an inter-governmental organization. Principle No. 12, therefore, is unlikely to be seen as problematic.

Principle No. 8 in the Joint Statement calls for "energy integration be-tween the countries of the region." This clearly refers to Petrocaribe and its sister companies, Petroandina and Petrosur. It carries no legally binding obligations to do anything, but it does bring up the question of the role of Venezuela vis-à-vis Trinidad and Tobago in the region's energy mix. Trini-dad and Tobago does not have the capacity to take over Venezuela's role as a supplier of crude oil to refineries in Caricom countries, and not much has been heard of late about its oil-financed Caricom aid facility. It seems unlikely that it could approach the scale of Petrocaribe's lending. On the other hand, there is no reason why the two sources should be regarded as competitors in the regional oil market in the present environment of tight energy supplies. Principle No. 8 may not conflict with Caricom obligations, since it is non-specific and non-binding, but it does point to the desirability of a coordinated Caricom energy policy.

Regarding Dominica's "statement of accession" to ALBA, this document is of a general nature. It recites ALBA's principles and achievements but contains no specific commitments or obligations by Dominica or any other ALBA country. It does not even go as far as the accession statement of Bolivia or the Cuba-Venezuela statement for the application of ALBA, which spell out certain trade and payment arrangements. Indeed, it does not refer to the Cuba-Venezuela Joint Statement creating ALBA or to any other document except the February 2007 joint declaration of the three OECS states. In short, there is no evidence in this document that Dominica has undertaken any specific obligations of any kind, let alone obligations that might conflict with those applying under the Revised Treaty of Chaguaramas or the CSME.

## Coordination of Trade Policy in Caricom

Article 80 of the Revised Treaty of Chaguaramas calls for the coordination of the external trade policies of Caricom member states. As such, it mandates the community to pursue the negotiation of external trade and economic agreements "on a joint basis." At the same time, it allows for the negotiation of bilateral agreements by member states, "in pursuance of their national strategic interests," as long as they are without prejudice to members' revised treaty obligations. Where they contain a trade or tariff component, bilateral agreements are subject to certification by the Caricom Secretariat or Council for Trade and Economic Development (COTED), as the case may be. Article 80 represents a compromise between regionalism and the preservation of national sovereignty in an area that is vital to the national interests of member states, given the highly trade-dependent nature of Caricom economies.

The ALBA documents signed by three Caricom countries do not contain a trade component, or any reference to tariffs. As such, they do not appear to require certification by the Caricom Secretariat or the COTED. The Petrocaribe agreements signed by thirteen of the fifteen Caricom members could be construed as having a trade component, especially where "compensated trade" is involved. When Caricom members signed on in 2005, concerns were expressed in some quarters[20] about the lack of prior Caricom consultation. This matter has never been resolved; in effect, Caricom members have agreed to differ over both ALBA and Petrocaribe.

## Economic Vulnerability Issues

Vulnerability relates to the issues of external indebtedness and possible political change in Venezuela. The type of concessional debt represented by Petrocaribe creates less debt-servicing obligations per dollar than commercial or IFI debt; nonetheless, the steep increase in total Petrocaribe debt should be a source of concern. Three strategies are called for here.

First, Caricom states should aim to keep, within a target ceiling, *aggregate* debt servicing obligations arising from all external debt as a proportion of exports (goods and services) and GDP. Petrocaribe debt should be managed as part of a total external debt management strategy.

Second, a high proportion of external debt owed to a single source opens the way to the exercise of political and economic leverage by the creditor. Here the strategy should be to regionalize relations with the creditor as far as possible while fixing a target ceiling for indebtedness to any one donor as a share of total external debt.

Third, Caricom states must as a matter of their long-term survival adopt aggressive policies for energy conservation and the development of new and renewable sources of energy, so as to reduce their reliance on special financing for energy imports. A portion of Petrocaribe could be set aside for this purpose. The availability of concessional loans to finance energy imports should not be allowed to reduce the incentive for energy conservation and development policies.

## Political Vulnerability Issues

The first issue that needs to be considered here is the matter of territorial claims and maritime boundary disputes involving Venezuela on the one hand and Guyana and Aves Rock on the other. There are concerns that the Caricom stand on these matters could be compromised by ALBA participation.

As things presently stand, there is no evidence that Venezuela has sought to establish a linkage, either formally or informally, between the issues and participation in ALBA/Petrocaribe. Certainly, ALBA and Petrocaribe documents make no reference to these subjects even in general terms, and political leaders and officials give no indication of informal linkages. Any attempt to establish linkage of this kind could be expected to elicit a firm and unambiguous response from Caricom states.

The second issue of political vulnerability is the possibility of change in leader, government, or political orientation in Venezuela. There is no way to assess the likelihood of such eventualities. President Chávez has been in office for the better part of ten years and has won several elections and referenda, losing only the last referendum by a narrow margin. His present term of office lasts until 2012, and he is not eligible for reelection; however, the Bolivarian mission appears to be a widely held ideology in Venezuela and may outlast the Chávez presidency. Certainly, disenchantment with neoliberalism and the Washington consensus has spread throughout much (though not all) of Latin America. It underscores the desirability of regionalizing relations with ALBA and Venezuela through joint or coordinated negotiations, and of strategies to mitigate debt dependence and energy dependence.

A particular point of concern is the possibility of incorporating a military dimension into ALBA and thereby being drawn into a military confrontation with the United States. These concerns arose from media reports on remarks made by the Venezuelan president in his speech at the Sixth ALBA Summit. This possibility, however, even if actually voiced by President Chávez (and this itself is not clear), was not supported by other leaders attending the Summit.[21] It was not mentioned in the Summit Declaration, nor is there any reference to military cooperation in any of the official ALBA documents. Officials of Caricom member states who deal with this matter insist that there is no discussion of a military dimension in ALBA. A more likely proposition is Brazil's proposal to establish a South American Defense Council under the umbrella of Unasur, a matter which is now the subject of discussions with two Caricom member states.[22]

## Relations with Other Hemispheric Powers

Observers have also raised concern about the potential effect of ALBA participation on relations with other hemispheric powers, notably the United States and Brazil. Given the poor state of Venezuela-U.S. relations and the nature of the ALBA mission, association with ALBA carries the risk of compromising Caricom's traditionally friendly relations with the United States, through "guilt by association." In the case of its relations with Cuba, Caricom has been able to maintain a clear distinction between the development of trade and cooperative relations with other hemispheric countries on the one hand, and support for its member government's statements and actions on the other. This is an essential attribute of the foreign policy of sovereign states on which Caricom has insisted in various arenas, and the same principle applies to relations with Venezuela and ALBA. The main areas of cooperation between Caricom and the United States—security, drug trafficking, money laundering, and migration—are all matters of mutual interest and should not be affected by participation in ALBA.

With regard to Brazil, there is the matter of its rivalry with Venezuela for influence in the LAC region. In this regard, participation in ALBA should actually increase the incentive for Brazil to deepen its cooperation with Caricom, as a means of countering Venezuela's influence.

## A Caricom-ALBA Agreement?

We have suggested that coordination of Caricom's trade policy with ALBA and Petrocaribe, as with other countries and trade groupings, is both a treaty obligation and desirable for economic and political reasons. A joint agreement, for example, could contain provisions designed to address Caricom concerns, such as respect for the provisions of the Revised Treaty of Chaguaramas and for the territorial integrity and sovereignty of member states. However, a coordinated position or joint agreement appears unlikely, because of differences among member states in the perceived costs and benefits of association with ALBA. Trinidad and Tobago continues to promote its candidacy as the site of the headquarters of the FTAA, to which ALBA is being promoted as an alternative. Other member states, such as Barbados and Jamaica, may have reservations about the impact of close association with ALBA on their relations with Washington. Both Trinidad and Tobago and Barbados have declined to participate in Petrocaribe, the former because, as an energy exporter, it considers Petrocaribe a competitor in the regional oil market.

This issue throws into relief the difficulty of coordination in a community with varying economic circumstances among member states. The difficulty was not as evident when Caricom's most important relationships were with traditional trading partners in the North Atlantic. Commonality of interest in trade relations with the EU, the U.S., and Canada made joint negotiations relatively easy. With global and hemispheric reconfiguration, the issue is more sharply posed. For example, Caricom is unable to adopt a coordinated trade policy with the People's Republic of China, since several member states maintain diplomatic relations with Taiwan. In addition, as was mentioned above, two Caricom member states on the South American continent are participating in the Union of South American States (Unasur) and in May 2008 signed the constitutive agreement formalizing the grouping.[23] As in the other cases, their participation was not the subject of prior community sanction or the result of a community-wide strategic policy. In effect, Caricom members have agreed to differ, not only on ALBA but also on several other subjects of external trade policy where they have divergent interests. The communiqué issued at the end of Caricom's Council for Foreign and Community Relations (COFCOR) meeting in May 2008, where ALBA was discussed, reflects this ongoing compromise.

## The Protection of Caricom's Strategic Interests in the Changing International and Hemispheric Context

Ministers examined the geopolitical and economic changes taking place at both the hemispheric and international levels and the resulting challenges. They considered these changes particularly in the context of the redistribution of power on the global stage; the shifting priorities of traditional partners; the increasing presence of non-traditional actors in the Caribbean region; and geopolitical changes in neighbouring regions and states. They also considered non-geopolitical issues having an impact on the Community including climate change, rising food and fuel costs and crime and security.

In this context, Ministers considered policy responses to these changes and challenges as well as new initiatives which have emerged and which could assist countries of the Region in meeting these challenges, among them the Bolivarian Alternative for the Peoples of our America (ALBA).

Ministers agreed that Member States should continue to pursue and explore all opportunities available to them for their social and economic development, recognising at all times their obligations under the Revised Treaty of Chaguaramas.

Extract from communiqué issued at the conclusion of the Eleventh Meeting of the Council for Foreign Aid and Community Relations (COFCOR), May 7–9, 2008, Bolans Village, Antigua and Barbuda. Caricom News Release, May 10, 2008, http://www.caricom.org/jsp/pressreleases/pre125_08.jsp.

In the absence of coordination, Caricom's external trade policy will continue to be a series of ad hoc bilateral responses to opportunities afforded by global and hemispheric reconfiguration. While this may be understandable in light of the community's divergent interests, the downside will be a failure to capitalize on the leverage available from coordination and the synergies of joint action. The fallout from external trade policy on domestic policy also brings the risk of regional fragmentation. The community could, in effect, be pulled in several different directions at the same time.

## Conclusion

This discussion argues that ALBA should be seen as one expression of a process of reconfiguration in world and hemispheric affairs. The ALBA model offers several attractive features from Caricom's point of view. These include flexibility in the terms of participation based on differentiated treatment, non-reciprocity in trade and payment arrangements, availability of considerable financial assistance on concessional and low-conditionality terms, and cooperation in health and education that directly benefits economically disadvantaged groups. Non-reciprocity should not be one-way, however; Caricom should consider establishing a development cooperation program in which it plays a donor role.

ALBA also benefits Caricom as part of a broader policy of strategic diversification in its external economic relations. Caricom could take advantage of ALBA's opportunities while attenuating its energy and donor dependency, preserving the integrity of its own integration arrangement, maintaining its political commitments to its own membership, and minimizing any negative effect on relations with its traditional partners. Success in achieving these objectives would be enhanced if Caricom adopted a coordinated position or negotiated a joint agreement with ALBA. At present, this does not appear likely, because of divergent national interests within the community.

The lack of a coordinated position within Caricom affects its relations with nontraditional trading partners in general. While the process of reconfiguration brings new opportunities for cooperation and strategic diversification, it also poses new challenges to the cohesion of the Caricom integration movement.

## Notes

1. U.S. Council on Foreign Relations, "Era of US Hegemony in Latin America Is Over, Says CFR Task Force," Press Release, May 15, 2008, www.cfr.org/publication/16245/era_of_us_hegemony_in_latin_america_is_over_says_cfr_task_force.html. The full report is published as "Report of an Independent Task Force, US-Latin America Relations: A New Direction for a New Reality," http://www.cfr.org/publication/16279.

2. Information on ALBA was drawn from the documents listed in Text Box 8.2 and from various documents at the Portal ALBA-TCP Website, www.alternativabolivariana.org, including "What Is ALBA?," at www.alternativabolivariana.org/pdf/alba_mice_en.pdf; Ministerio de Integración y Comercio Exterior, "El ALBA en el Caribe," at www.alternativabo-

livariana.org/modules.php?name=Content&pa=showpage&pid=258; and Venezuelan Bank of External Commerce (BANCOEX), "What Is the Bolivarian Alternative for Latin America and the Caribbean?" February 5, 2004, at www.venezuelanalysis.com.

3. Besides the above-named seven countries, the leaders of Ecuador, St. Kitts-Nevis, Honduras, Haiti, and Uruguay were also participants.

4. Acuerdo de Cooperación Energética Petrocaribe, www.alternativabolivariana.org/modules.php?name=Content&pa=showpage&pid=232.

5. Acuerdo entre el presidente de la República Bolivariana de Venezuela y el presidente del Consejo de Estado de Cuba, para la creación de la Alternativa Bolivariana para las Américas, Article 11, December 14, 2004, http://www.alternativabolivariana.org/modules.php?name=Content&pa=showpage&pid=2060.

6. Acuerdo para la aplicación de la Alternativa Bolivariana para los pueblos de nuestra América y el Tratado de Comercio de los Pueblos, April 29, 2006, www.alternativabolivariana.org/modules.php?name=Content&pa=showpage&pid=516.

7. Acuerdo entre el presidente de la República Bolivariana de Venezuela y el presidente del Consejo de Estado de Cuba, Article 18, http://www.alternativabolivariana.org/modules.php?name=Content&pa=showpage&pid=2060.

8. Acuerdo para la aplicación de la Alternativa Bolivariana para los pueblos de nuestra América, www.alternativabolivariana.org/modules.php?name=Content&pa=showpage&pid=516.

9. Acuerdo entre el presidente de la República Bolivariana de Venezuela y el presidente del Consejo de Estado de Cuba, Article 8, http://www.alternativabolivariana.org/modules.php?name=Content&pa=showpage&pid=2060.

10. "Dominica Makes History as ALBA Member," *Caribbean World News*, January 15, 2008, www.caribbeanworldnews.com/middle_top_news_detail.php?mid=136.

11. Havelock Brewster, "Understanding Development Challenges in the Caribbean: Time to Take in the Begging Bowl," May 2007, www.normangirvan.info/understanding-development-challenges-in-the-caribbean/.

12. Nidia Diaz, "Venezuela Offers to Finance 50% of ALBA Nations' Oil, *Granma International*, May 4, 2007, at www.granma.cu.

13. "Dominica Makes History as ALBA Member," *Caribbean World News*, January 15, 2008, www.caribbeanworldnews.com/middle_top_news_detail.php?mid=136.

14. Information on Petrocaribe was obtained from the following sources: PetroCaribe Summit Website, www.jis.gov.jm/special_sections/summit/; Petróleos de Venezuela, S.A., www.pdvsa.com/index.php?tpl=interface.en/design/readmenuprinc.tpl.html&newsid_temas=48; Ministerio del Poder Popular para la Energía y Petróleo Petrocaribe, www.mem.gob.ve/Petrocaribe/index.php; Kaia Lai, "Petrocaribe: Chávez's Venturesome Solution to the Caribbean Oil Crisis," at www.venezuelanalysis.com.

15. The figure of $4.5 billion by 2010 may turn out to be a considerable underestimate, as it was given in December 2007, before the huge increase in oil prices in the first half of 2008.

16. Mark Sullivan, "Caribbean Region: Issues in US Relations," Congressional Research Service report to Congress, October 27, 2006.

17. In other words, between 25 percent and 50 percent of the extra cost of 24,000 barrels per day (8.76 million barrels per year) at a price of $120. This figure includes Barbados's imports, but Barbados opted not to participate in Petrocaribe.

18. Sistema Económico Latinoamericano y del Caribe, "El Banco del Sur comenzará a operar a finales de 2008 con un capital inicial de 7.000 millones de dólares," www.sela.org/sela/prensa.asp?id=13115&step=3.

19. Venezuelan Embassy in France, "Tema alimentario será prioridad en Cumbre del ALBA," April 23, 2008, www.embavenez-paris.com/embavenez.php?cat=pei&inc=24_04_2008.

20. Specifically, by Prime Minister Arthur of Barbados and Prime Minister Manning of Trinidad and Tobago.

21. Jossette Altmann, "The ALBA Bloc: An Alternative Project for Latin America?" April 17, 2008, www.realinstitutoelcano.org/wps/portal/rielcano/contenido?WCM_GLOBAL_CONTEXT=/Elcano_in/Zonas_in/ARI17-2008).

22. "Consejo de Defensa Suramericano profundizará la integración," www.alternativabolivariana.org/modules.php?name=News&file=article&sid=3000.

23. Tratado Constitutivo de la Unión de Naciones Suramericanas (Unasur), www.alternativabolivariana.org/modules.php?name=News&file=article&sid=3010.

9

# European Progressives
# and the Bolivarian Social Agenda

JULIA BUXTON

The following discussion addresses "progressive" European perspectives of Hugo Chávez and his Bolivarian revolution. Within this framework, the aim is to outline the drivers of Chávez's popularity among left-of-center and so-cial-activist groups in Europe, and to explore whether identification with the *chavista* phenomenon is a result of ignorance by distance or something more positive. Put simply: Are progressive Europeans unable to analyze Venezue-lan developments objectively because of their tendency to romanticize South America's revolutionary tradition? Does subjectivism blind those across the Atlantic to a regressive and authoritarian reality?

We begin with an overview of European engagement with South America as a means of providing context for the discussion, continuing with an out-line of the policies and factors that have led progressive European actors to become engaged with Bolivarianism and the domestic and foreign policy of the Chávez administration. Two stages of progressive alignment with the Bolivarian process are identified, the first commencing with the failed coup attempt of April 2002, and the second dating roughly from the recall refer-endum of 2004. To avoid sweeping generalizations, the discussion focuses specifically on Western European groups, in the United Kingdom in particu-lar. This emphasis on the British experience owes not so much to the author's Englishness but rather to the vibrancy of activism around Venezuela in the U.K. More than forty Members of Parliament belong to the Labor Friends of Venezuela; successive conventions of the Trades Union Conference have passed resolutions supporting Venezuela's progress in addressing poverty, and the leading trade unions are affiliated with and fund pro-Venezuela campaigns; and the country is home to two active information and solidarity campaigns. The Scottish National Party, which controls the devolved Scottish Executive, is also sympathetic to the Chávez government, as are a number of unions and Republican councilors in Northern Ireland.

We continue with a discussion of the factors that influence European interest in constructively debating the social dimensions of Chávez's policies, among them Europe's high profile in donor and development aid (placing debate on Venezuela within a wider discourse on anti-poverty interventions); the socialist tradition and vibrancy of socialist parties in European politics, as well as a tradition of social welfare (leading to more support for state-led anti-poverty initiatives and non-neoliberal alternatives); and the European academic tradition, with its emphasis on qualitative methodological approaches. In shifting beyond number-crunching, quantitative critiques with a zero-sum view of the success of Chávez's social agenda, the European debate addresses positive as well as problematic factors, moving beyond the polarization that characterizes the North American discourse and allowing for constructive criticism. Finally, we will consider the extent to which Chávez's defeat in the December 2007 constitutional referendum and revelations of his alleged links to the FARC may have damaged support and sympathy for the Venezuelan government among European progressives.

## The Bigger Picture: Europe and South America

To be simplistic, South America is seen from Europe as the North American backyard. European institutions are largely excluded from geographically defined hemispheric debates on trade, security, and immigration, interposing only through the framework of bloc-to-bloc ties. While there are examples of EU bloc-to-country relations (e.g., with Mexico and Chile) and bilateral country-to-country ties, South America in general is defined as in the United States' sphere of interest.

European engagement with South America is structured primarily around commerce and trade. The EU also has a social agenda in the region; however, taken together, individual European member states and the EU itself are the largest development donors in the hemisphere. European cooperation initiatives are framed by Article 177 of the EC's founding treaty and the 2005 European Consensus on Development, which base community policy on the sustainable economic and social development of developing countries, support for their integration into the global economy, poverty alleviation (in line with the Millennium Development Goals), and support for good governance and the rule of law. Specifically with regard to the Andean region, EU dialogue is based on the 1996 Rome Declaration, which was superseded in 2003 by the EU-Andean Community Declaration. These guidelines allow bloc-to-bloc discussion of issues such as terrorism, conflict prevention, and good governance, set within the broader commerce and trade-oriented EU-Andean Cooperation Agreement that came into force in 1998.

The multidimensional nature of EU relations with South America, and the Andes in particular, is likely to become more tightly focused as a result of the 2003 Common European Security Strategy.[1] This strategy identifies five key threats to European integrity and stability: terrorism, organized crime, failed states, regional conflicts, and proliferation of weapons of mass destruction. Action on these fronts necessitates enhanced cooperation with South American countries, specifically with regard to drugs and organized crime.

A few contextual observations are important before we move on to address progressive actors and movements in Europe and the sympathy that the Bolivarian process has engendered among these groups. First, the EU maintains a multilateral, multinational security framework that is agreed and upheld across all member states. From the progressive European perspective, the United States' often unilateralist stance in South and Central America, working outside of the OAS, is an anathema. Second, European integration is premised on and facilitated by financial cross-transfers from wealthier European states to poorer ones. Greece, Spain, and Ireland in particular are historic beneficiaries of this initiative, and the new accession states of Central and Eastern Europe are replicating this experience. In this context, initiatives such as the ALBA are not perceived as problematic; rather, it is the "hub-and-spoke" model of integration in the Americas, which has been devoid of supranational institutions, civil society oversight, or cross-national subsidization, which runs counter to the European "neighborhood" experience. Moreover, while integration in the Americas has been devoid of social measures, European countries have extended their long tradition of state welfare and social protection into supranational policy and have zealously guarded the rights of unions.

One other area of divergence in the EU/Americas experience is significant for understanding progressive perceptions of the Bolivarian process and the conclusion that revelations of alleged collusion with the FARC are unlikely to cost the Venezuelan government support among its European sympathizers. In contrast to the United States' militarized response to the problem of narcotic drugs and protracted insurgency in Colombia, the EU has maintained that these security challenges need to be addressed by alternative development and dialogue, with emphasis on the structural drivers of the conflict.

## The EU and Chávez: Not a Love Match

While the EU considers itself a progressive actor in South America, this has not translated into support for Hugo Chávez or his Bolivarian revolution at the EU intergovernmental level. The essence of Bolivarianism runs against the EU strategy for commercial engagement in the region, and protagonistic democracy and the organizational principles of twenty-first-century socialism are incompatible with European ideals of liberal democracy and the rule of law. In practice and policy, the Chávez administration has countered the fundamentals of EU engagement with South America, and his government is seen to have acted against the interests of a substantial number of European transnational companies (in mining, finance, agriculture, oil, and gas). Venezuela's withdrawal from the Community of Andean Nations, the founding of ALBA, its rejection of free trade, and its nationalization strategy have been major setbacks for the EU framework. In sum, the European Union and its institutions should not be counted among European progressives aligned with Bolivarianism.

Chávez's ascent to power and the evolution of his administration have also met with a jaundiced response from the individual national governments of many EU member states. This position has been reinforced by Chávez's unusual and quite vocal diplomacy, as experienced by Tony Blair (a "poodle"), the king of Spain, and, more recently, German chancellor Angela Merkel (of "Fascist" Hitler lineage). Perceived threats to the commercial interests of individual European countries have been a primary cause of the disaffection of many European governments with the Chávez administration, as have Bolivarian social policies. Europe (and member state donor agencies[2]) promotes conditional cash transfer programs and global integration as the mainstay of poverty reduction, a policy position that sits uneasily with Chávez's emphasis on the redistribution of economic as well as political power. EU member governments tend to follow the "good" left and "bad" left dichotomy that characterizes U.S. efforts to simplify complex and nationally driven trends across the Southern Hemisphere. The democratic characteristics of the Venezuelan government have been questioned, and former Eastern Bloc countries—specifically, the Czech Republic and Poland—have sought EU measures against Venezuela's "populist authoritarianism."[3] In stark contrast to the United States (and the Czechs), however, European governments and the EU itself have refrained from sharply combative language and have consistently pursued national and regional policies of dialogue with Venezuela (and Cuba).

Developments in Venezuela have been an issue for some European governments more than others, of course. Spain, France, and the UK have, for a variety of commercial and historical reasons, been more "engaged" with Venezuela than administrations in, for example, Austria, Sweden, or the Netherlands, and this discrepancy is mirrored at the grassroots level among progressives. Even within proactive countries, policy and position have shifted with changes of government, and engagement with Venezuela has not always been benign. Senior figures within the Venezuelan government implicated the UK and Spanish administrations of former Prime Ministers Tony Blair and José María Aznar in the April 2002 coup attempt against Chávez—an event that was greeted with delight at the time by former British Foreign Office Minister Dennis MacShane.[4] These tensions in turn underpin the mobilization of progressives around the Venezuelan "cause." European critics of British and Spanish policy during this period, including the Labor Friends of Venezuela in the British Parliament, believe that both countries were acting under the influence of the United States. In turn, where there has been more national government engagement (good or bad) with Venezuela, there has been more engagement by progressives. Developments, division, and political debate in Europe during this period are instrumental to understanding the alignment of progressive Europeans with the *chavista* cause. In many cases, it sets progressives at odds with their national governments, and reflects the fact that support for Chávez tends to come from a minority within the complex of political actors and institutions.

By way of an introduction to the following section, we must emphasize that European engagement with, and interest in, South America cannot be explained solely through government or regional bloc ties. Spain, of course, has a long history in the region, but it is in North America that we find the bulk of scholarship on the region, and South American issues are, for obvious reasons, more visible in the North American media. By contrast, the study of South America appears to be undergoing something of a decline in Europe. Many European countries have no knowledge base or institutional source of expertise on South America, and media coverage tends to be minimal, supercilious, or simply vacuous—pandering to stereotypes of banana republics awash with poverty, narcotics, guns, and corruption.

Despite these limitations, South American politics has been a sustained focus of interest for progressive actors and movements across Europe. It is worth noting that, on the whole, they have received no support, or even interest, from the Venezuelan government. Venezuelan diplomacy has been largely noticeable by its absence in Europe—a significant foreign policy limitation in the view of this author, and one that contrasts strongly with the political "activism" of the Venezuelan Embassy in Washington, D.C.

## Progressive Europeans: Actors and Sympathies

### Phase 1: Courtship and Romance

If the Clinton administration viewed Chávez's election with misgivings, European governments seemed as yet not attuned to the political perils posed by the incoming president. For the Americans, Chávez's victory meant watching an individual whom Secretary of State Madeline Albright had called a terrorist assume power in a strategic oil-supplying country. European governments, in contrast, greeted (albeit with some trepidation) a man who cited Tony Blair as an inspiration and third-way socialism as a model.[5] Among European progressives, Chávez's ascent to power registered limited to no interest. Trade union organizations, student movements, and the political Left had little time for a government that professed no significant or revolutionary platform beyond rewriting Venezuela's constitution and whose radical credentials at this point rested simply on overturning an old and sclerotic two-party system.

The spark that catalyzed the first phase of solidarity with Venezuela was the coup attempt of 2002. This event, continuing with the PDVSA stoppage and lock out and the recall referendum of August 2004, was decisive for European progressives' engagement with Venezuelan politics and mobilization around the Bolivarian cause. Three factors were of particular significance. First was the generalized disdain among "liberal" West European opinion of the opposition's actions against Chávez. Carmona's move to dismiss all elected officials and to rule by decree through a junta recalled the dark days of right-wing authoritarianism in South America. At this point, the democratic credentials of the Chávez government were seen to be sound, and his program of government relatively uncontroversial. European activists and journalists, and academics[6] with experience in Chile and Nicaragua were drawn to defend Venezuela on the basis that history was, in effect, repeating itself.

A second and interlinked factor was the United States' role in the coup, which fueled negative sentiment against U.S. intervention in South America and in world politics more generally. Again, historical experience and knowledge of U.S. intervention and collusion with undemocratic right-wing forces attracted progressives to the Chávez government. Venezuela was compared to Iraq as an example of oil-induced motivations for U.S. actions and, coming at the same time as the U.S.-led invasion of Afghanistan, of a self-interested, militarized U.S. foreign policy that violated international law, principles of state sovereignty and—in the case of Venezuela—democracy. In this context, the failure of the British government to condemn the coup attempt against Chávez was seen as further evidence of Tony Blair's role as a "poodle" to the Bush administration, and inspired the creation of the Labor Friends of Venezuela group.[7]

The third important catalyst of progressive alignment with the Chávez government was coverage of the events of 2002 by the European, North American, and Venezuelan media. Two issues were important here: the remarkably unsympathetic depiction of Chávez, and a failure to contextualize and analyze events leading up to the coup attempt (and its subsequent collapse). Among European media, right-of-center publications, such as the *Financial Times*, the *Times*, *El País*, and particularly the *Economist*, stand out for their failure to grasp the salience of Chávez's appeal and for the overtly partisan nature of their coverage. Even respected left-of-center news outlets, such as the *Guardian* and the BBC, failed to provide an effective voice for the deposed (and then resurrected) *chavistas*. The poor and subjective nature of the coverage fueled broader concern among progressives over the decline of the journalistic tradition, the depletion of foreign correspondents, the dominance of multinational media corporations, and overreliance on newswires for information. Given the strongly anti-Chávez tone in the North American media and revelations of the role of the Venezuelan private media in the coup attempt, European progressives were drawn to the need for balanced information that countered what they saw as a deliberate and controlled campaign of disinformation. In the UK, they condemned negative media coverage, by the *Guardian* in particular, as being directly related to the paper's affinity with Tony Blair and his "New Labor" project. This disaffection reinforced a growing division within the British Left around Venezuela that was replicated in a number of other West European countries. The Irish television documentary *The Revolution Will Not Be Televised* proved to be a particularly important tool for generating progressive interest in Venezuela and debate on the role of the Venezuelan media.

In response to these concerns, progressives formed two organizations that went on to acquire national and European momentum: Hands Off Venezuela, a Trotskyite organization headed by Alan Woods; and the Venezuelan Information Centre (VIC), founded by a group of academics, journalists, antiwar groups, and trade unionists, with London Mayor Ken Livingston as patron. The issue in question was not the need to promote a *chavista* perspective (which many members of these groups would in any event have been hard-pressed to identify at this stage), but to ensure accurate and truthful reporting. The broader context was concern over the role of the media in "spinning" information and (underscoring the importance of chronology) its failure to critically investigate claims of weapons of mass destruction in Iraq.

Phase 2: Marriage

The second phase of progressive engagement with Venezuela dates from the period of the recall referendum campaign. During this phase, a growing number of people were drawn to the Venezuelan cause, and solidarity campaigns became more coherent and organized. Academic, media, and policy interest in Venezuela deepened, as did divisions between those supportive of the government's record and those critical of it, particularly on the left.

During this period, the Chávez government evolved a more radical ideology and rolled out policies and positions with which European progressives could more readily identify. Rising oil prices, the need to consolidate support among the poor, the unveiling of regional integration initiatives designed to insulate Venezuela from the U.S., and the purges that followed actions by the opposition[8] gave the Chávez government a more coherent identity. The government initiatives that followed—the Missions, ALBA, Telesur, Banco del Sur, *cogestión*, Banco de la Mujer, Community Councils, MTAs—were deeply interesting to progressive European actors, who saw them as an experiment in building an alternative to neoliberalism, free trade, and procedural liberal democracy. Their attraction was neither blind nor uncritical; on the contrary, internal debate among progressives was vibrant and at times conflictive. There was a consensus, however, that, regardless of its policies, Chávez's government was elected and supported by poor and marginalized Venezuelans. The progressive campaign, therefore, was built around the right of the Venezuelan people to be served (ill or otherwise) by the government they had elected.

The following section summarizes some of the Venezuelan government's policy initiatives and why they were of interest to progressive groups.

Domestic Policies

*The Missions.* The government's expansive social policy agenda on education, health, nutrition, employment training, land distribution, and credit availability was intellectually attractive to progressives for a number of reasons. First, and by way of context, progressive groups were increasingly frustrated at the lack of progress that donor agencies, such as DFID and UNDP, were making toward achievement of the Millennium Development Goals. An anti-poverty agenda premised on conditional cash transfers, poverty reduction strategy papers, public-private initiatives, global integration, and free trade did not appear to be delivering the economic growth benefits required to alleviate poverty. The Missions were interesting on a number of counts:

1. They represented a domestically developed anti-poverty initiative with the potential to galvanize domestic legitimacy, maintain credibility among the poor, and engage stakeholders—the big catchword of the donor community.
2. The Missions were unusual in their aim of redistributing political power as well as economic power, representing an important break with the donor tradition of focusing purely on the economic drivers of poverty and exclusion.
3. The Missions were organic and holistic in their conceptualization, with a synergy in basic social-service provision. They recalled the welfare model and cradle-to-grave support of the Western tradition (although delivered in a manifestly different context of informal,

parallel organizational structures). As such, they broke with the model of conditional cash transfers, which progressives critiqued for addressing only one minor element of the poverty trap. The Missions were also accompanied by policies to redistribute land ownership, to mainstream gender provision, and to address rural/urban inequities—strategies that are recognized as fundamental for sustainable livelihoods and poverty reduction but which are rarely borne out in practice.

4. The Missions specifically targeted and disproportionately benefited the poor (leading critics to condemn the policies as clientelist and a zero-sum game) in an attempt to break the intergenerational transmission of poverty. And in promoting income growth for the poor *in addition to* human capital development, the Venezuelan government embraced a dual policy strategy that is recognized as fundamental for the elimination of poverty in poor and unequal societies but that, again, is rarely put into practice.

*Civil, Social, and Workers' Rights.* The Venezuelan government's promotion of the social, civic, organizational, human, cultural, and economic rights of women, children, gays and lesbians, indigenous peoples, workers, and Afro-Venezuelans also made the Bolivarian process attractive to progressives. Although liberals and critics of the government inevitably (and legitimately) questioned whether this goal is achievable within a framework characterized by a weak rule of law, for progressives the value of the Bolivarian process rested in constructing a dialogue—a narrative and legal framework within which these rights are recognized and through which they can be articulated and defended. The promotion of a "native" and historical cultural identity also found resonance among groups disaffected with the culturally homogenizing thrust of globalization. In sum, the Venezuelan government was seen to be addressing issues traditionally off the policy agenda of national governments, while at the same time overturning a legacy of patriarchy, elitism, and (unacknowledged) racism.[9] The promotion of workers' rights was also seen to be a positive step in a region characterized by repressive labor policies.

*Statism.* In contrast to U.S. critics, European progressives did not automatically reject the Venezuelan government's nationalization strategy. The promotion of the state in the national economy was not specifically controversial in a European context traditionally characterized by high levels of state intervention and deepening public misgiving over the privatization of industries, utilities, and welfare in many Western European countries in the 1980s and 1990s. Outside of the immediate European context, progressives were deeply critical of multinational corporations and of unfair terms of trade in the world economy. A generic hostility to privatization was conjoined with a critique of the neoliberal legacy in South America, resulting in support for and interest in the Venezuelan government's approach—or, at a minimum, respect for its sovereign right to conduct economic affairs of state.

*Anti-Neoliberalism.* Chávez's somewhat late affinity with the anti-neoliberal cause (which was only forcefully articulated at the World Social Forum in 2005) is perhaps one of the most attractive elements of his administration for progressives. A substantive body of criticism within the European academic mainstream is devoted to the negative and profoundly regressive impacts of neoliberalism in the Americas. This school of thought associates neoliberalism with deepening economic inequality, social injustice, and political conflict. In rejecting neoliberal policy measures, and in his stirring critiques of the International Monetary Fund and World Bank, Chávez struck a deep chord among critics of neoliberalism, of the international debt of developing countries, and of multilateral institutions.

## Radical Democracy

A final point of interest for progressives that merits attention here was the efforts by the Chávez government to promote participatory democracy and its community councils experiment. European progressives supported the sovereign and electoral right of Venezuela to develop its own institutional model and to break with procedural models of liberal democracy condemned as unrepresentative and elitist. From a development perspective, the community councils were an interesting experiment in direct and stakeholder participation and sparked scholarly interest in the mechanics and experience of the participatory process.

On a slightly different level, but again revealing the drivers of progressive identification with Bolivarianism, developments in Venezuela confirmed existing concerns with the role of U.S. democracy-promotion agencies such as the National Endowment for Democracy. Even academics scornful of the Chávez administration saw the country as a test case of all that had gone wrong with the global democracy promotion and liberal peace agenda.

Foreign Policy

As the United States considered placing Venezuela on its list of state sponsors of terrorism, decertifying the country for noncompliance in the war on
drugs, and maintaining an arms embargo on the country, European progressives placed themselves solidly behind the Chávez administration on a
number of issues that are a cause of consternation and insecurity in the U.S.
At the government level, the EU and national European governments see the
current U.S. approach as undermining prospects for dialogue, a possibility
they maintain should be kept open at all times.[10]

  *Cuba*: The United States' position on Cuba is something of a conundrum,
and an anathema for Europeans. At both the grassroots and governmental
levels in Europe, the U.S. embargo is seen as a profound failure, entrenching
the Castro regime and beholden to the interests of a narrow and sectarian
lobby with political clout above and beyond its weight. The vibrant Cuban
solidarity campaign across Europe is motivated less by the iconic allure of
Che Guevara T-shirts and the romantic revolutionary tradition than by the
perception that the embargo is cruel, counterproductive, and illegal. Most
Europeans do not consider Chávez's close relationship with Cuba to be in
any way problematic. In this context, the Bush administration's claim that
Venezuela and Cuba constitute some form of axis of evil has been greeted
with derision in many European quarters.

  *Anti-Imperialism*. Rather than anti-American, Chávez should perhaps be
seen more rightly as anti-Bush within the broader context of a Venezuelan
foreign policy agenda that has consistently maintained strong anti-imperialist tones—from Bolívar to Cipriano Castro, from Luis Herrera Campins
to Carlos Andrés Pérez. Chávez's anti-Bush rhetoric has been an important
source of attraction for European progressives to the Bolivarian cause—if
anything, it could be said that given the depths of anti-Americanism across
Europe during the Bush years, Chávez may actually have galvanized more
support by escalating this rhetoric. Moreover, his attempts to diversify economic and commercial relations away from the U.S. were seen in Europe as
a rational, if not inevitable response to the changing realities of the global
political economy.

  *The Middle East*: On the one hand, Venezuela's alignment with Iran is
problematic for progressives, due in part to the nuclear technology issue and,
at a minimum, because of the issues of women's and human rights. On the

other hand, recent revelations that Israel owns nuclear warheads substantiated Chávez's claims of hypocrisy in the nuclear debate. Moreover, Western Europe is home to many activists for Palestinian causes, and Chávez's aggressive critique of Israeli/U.S. policy in the region has brought many progressives into the fold.

*Colombia*: A question posed at the beginning of this chapter is the extent to which alleged links to the FARC will undermine support for Chávez among European progressives. The answer is that they are unlikely to do so. Colombia is a major issue for European activist groups, and the vibrant and diverse Colombian solidarity campaign includes a large number of Colombian expatriates. Of course, solidarity should not be conflated with support for the FARC, just as support for a peaceful resolution of the Israel-Palestine conflict, or of the Cuba-U.S. dispute, does not mean support for Hamas, or for Fidel Castro.

The EU position has been one of promoting peaceful dialogue in Colombia—a position that stems as much from pressure by civil society organizations as from the EU predisposition to view militarized responses as counterproductive. Chávez's engagement in humanitarian exchange negotiations has been welcomed by many Western European governments, and positively encouraged by the French, and in this context linkages between Chávez and the FARC are accepted as inevitable. A second, related point is that many Europeans, progressive and conservative, believe the term "terrorism" is not conducive to peace and political stability. In the UK, peace came to Northern Ireland and mainland Britain through dialogue, negotiation, and a recognition of the underlying drivers of the conflict. Dropping the label "terrorism" was fundamental for progress in resolving "the Troubles." In this context, North American rhetoric emphasizing Chávez's collusion with "terrorists" does not have traction in Europe, where the EU decision to place the FARC on its list of terrorist organizations has been called into question.

Many European organizations (human rights, development, solidarity, and religious) are also disaffected with the current EU position on Colombia. These groups are pressuring the EU to resist intense lobbying by the Colombian authorities and to focus instead on European objectives in the country: a negotiated solution to the conflict; structural changes that address the drivers of impunity and conflict; and civil society incorporation into policy deliberation and delivery. In particular, the coalition of interests around Colombia is demanding that the EU base financial support and future cooperation on the country's compliance with UNHRC recommendations, human rights, and international humanitarian law, determined through transparent indicators. The same groups agree that Venezuela has a right to be involved in the resolution of the Colombian conflict, based on the belief that neighborhoods matter in conflict resolution, and on Venezuela's long history of dialogue with all actors in the Colombian conflict.[11] Colombia's insurgency is a security issue for Venezuela, and from a European perspective (in the context of multilateral European security strategies) it is fundamentally wrong for the U.S. to define the contours of and responses to the insurgency.

*Narcotics:* A final issue area that draws progressives to Bolivarianism and highlights the differences between the European and U.S. perspectives relates to narcotics. Many European groups supported Chávez's decisions to prohibit U.S. narcotics surveillance flights over Venezuelan territory, and to expel DEA agents from the country. Similarly, Chávez's support for the legalization of coca, his government's emphasis on demand rather than supply-focused responses, and criticism of Plan Colombia and U.S. counter-narcotics efforts, finds echoes in the EU and European academic literature. In Europe, the U.S. move to decertify Venezuela from the war on drugs was seen as counterproductive, by progressives as well as national governments. Europeans—like Venezuelans—find themselves on the raw receiving end of U.S. counternarcotics initiatives, such as Plan Colombia and the Mérida Initiative, which have diverted trafficking routes away from the Central American and Caribbean corridor, to Venezuela, Brazil, West Africa, and into Europe. As such, European progressives view Venezuela's "new" role in the drug trade as being by default rather than by design, and they seek to address this grave security threat multilaterally rather than through sanction and bombast.

A second and interrelated cause of disjuncture between Europe and North America relates to the elision of the war on drugs and the war on terror. As has become clear in debates on Afghanistan and Colombia, Europeans regard this elision as a fundamental policy mistake that conflates very different operations and phenomena at the risk of deepening these two separate problem areas.[12] Finally, there is a growing consensus in Europe that solutions to the challenge of illicit drugs must be devised regionally and multilaterally.

## Progressive Solidarity

A range of European groups and actors, variously labeled progressive, have come to associate with the Bolivarian process. This engagement is driven by a variety of factors and interests, but is underpinned by (a) a shared critique of global and international politics; and (b) support for Venezuelan sovereignty in a hemisphere dominated by U.S. unilateralism. The coalition includes journalists from *Le Monde Diplomatique* and Memoir de Luttes; individual senators and congressional representatives from the French, Belgian, Greek, Italian, and Spanish Socialist parties; the British Labor Party and the German Der Linke Group; student and labor movements from a range of European countries; antiwar groups, lesbian and gay organizations, black activists, women's organizations, and academics and writers. They constitute a broad and eclectic alliance, with an ideological spectrum that ranges from Social Democrats to Trotskyites. Their links to Venezuela have been strengthened by visits to the country and visits of Venezuelans to Europe. The Venezuelan government and its Diplomatic Service have been largely marginal to this process, with contacts forged more on a sector-to-sector basis than through any grand strategy. Infrastructural and logistical support has been provided by some but not all embassies (usually in the form of rooms for events, or the provision of speakers), and the funding of activities has been left to the campaigns themselves.

It would be wrong to view all progressives as pro-Venezuelan. Some of the harshest critics of the Bolivarian process come from the European Left and center-Left, and from those who were once on the radical left but who have been reborn on the right.[13] European progressives are not blind to the limitations of the Bolivarian process, and their support is not unconditional. Setbacks for the Venezuelan government, such as the defeat of the constitutional reform process in December 2007, are respected, on the basis that they represent the voice of the Venezuelan people. The solidarity European progressives express with Venezuela is solidarity with ordinary citizens as much as it is with the country's elected government. For progressives, Venezuela is an interesting experiment in constructing an alternative to the current global status quo.

## Notes

1. See A. Bailes, "The European Security Strategy: An Evolutionary History," SIPRI Policy Paper No. 10, Stockholm International Peace Research Institute, 2005.

2. Such as the United Kingdom's Department for International Development (DFID).

3. See Council on Hemispheric Affairs, "The EU and Colombia: Betraying Responsibility," November 10, 2005, at www.coha.org.

4. It was a matter of some embarrassment for Mr. MacShane that Chávez was restored to power in only a matter of days. In an interview shortly after the coup, MacShane described Chávez as "a ranting populist demagogue" who reminded him of Mussolini. In 2008, under recurrent attack (and mockery) for his remarks, MacShane denied that he had ever made these comments—which are available in the archives of *The Guardian* and the *Times* online.

5. See Richard Gott, *In the Shadow of the Liberator* (London: Verso, 2003).

6. Including, in the UK, social activist Bianca Jagger; playwright Harold Pinter; filmmaker John Pilger; journalists Richard Gott, Tariq Ali, and Hugh O'Shaughnessy; and the former mayor of London, Ken Livingston.

7. The leaders were MP Jeremy Corbyn, a veteran Latin American activist, and MP Colin Burgon, who is associated with the Cuban Solidarity Campaign and the Justice for Colombia campaign.

8. The coup attempt and PDVSA lockout enabled the government to remove disaffected elements in the armed forces and national oil industry, while work stoppages, capital flight, and nonpayment of taxes provided a pretext for the government to expand its role in the national economy, through, for example, currency and exchange controls.

9. See Barry Cannon, "Class/Race Polarization in Venezuela and the Electoral Success of Hugo Chávez: A Break from the Past or the Song Remains the Same?" *Third World Quarterly* 29, no. 4 (2008): 731-48.

10. It should be stressed that progressives too have points of division with the Chávez government, in particular relating to Venezuela's positions on Darfur and Tibet.

11. See S. Angeleri, *Guerrillas y búsqueda de paz en Colombia* (El Centauro: Caracas, 2000).

12. See, e.g., S. Cornell, "The Interaction of Narcotics and Conflict," *Journal of Peace Research* 42, no. 6 (2005); and "Crop Spraying: A Déjà Vu Debate—From the Andean Strategy to the Afghan Strategy," TNI Drug Policy Briefing Paper, No. 25, 2007.

13. E.g., Michael Reid, Latin America editor of *The Economist*; Richard Lapper of the *Financial Times*; and Isabel Hilton.

# 10

# How to Fill a Vacuum

Chávez in the International Arena

JORGE G. CASTAÑEDA

Latin America today is a region of contradictions. Economically, the region as a whole has been enjoying its highest economic growth rates in three decades, thanks to sensible macroeconomic policies and high prices for commodities—copper, iron ore, soy, and hydrocarbons. At least until recently, this growth has been accomplished without inflation or internal imbalances. There are undoubted weaknesses—for example, a possible drop in raw materials prices, or the consequences of a prolonged recession in the United States—but the economic surge is real and significant. It has generated an undeniable reduction in poverty, a small but promising leveling of inequality, and, above all, a notable expansion of the lower-middle class, even in countries as historically unequal as Mexico and Brazil. Politically, the scenario is equally positive. The transitions to democracy of the 1980s have turned out to be lasting and profound. Except for the Cuban regime and the FARC in Colombia, all of the region's political actors compete for power at the ballot box: even Hugo Chávez accepts his electoral defeats. Respect for human rights, while imperfect, is more established than ever before, and complaints are handled effectively and with accountability. Latin America is perhaps less important in the world than it was in the past, but to a large degree this is because it generates fewer problems.

Why then is the region caught up in more diplomatic, political, and social conflicts than at any other time in recent history? Starting north and moving south, in Mexico we find extreme political polarization that has paralyzed the government since Felipe Calderón took office. Colombia is involved in a series of disputes with its neighbors—with Nicaragua over the island of San Andrés and maritime limits; growing animosity with Venezuela; disputes with Ecuador over border violations and environmental issues—as well as its ongoing internal armed conflict. Peru has clashed with Venezuela over the "Casas del ALBA" (centers developed to spread Bolivarian revolutionary ideology), and with Chile over maritime limits. Bolivia has seen its fragile central government threatened by secessionist tensions between the country's Eastern Provinces and the Altiplano; has argued with Argentina and Brazil over energy exports; and has continued its historic dispute with Chile over access to the sea. Argentina has endured a prolonged agricultural strike and social polarization, as well as an impasse with Uruguay over pollution from a paper mill on the border between these two countries.

## A Polarized Continent

Analysts suggest many different explanations for the root causes of these conflicts, among them, the effects of narcotics trafficking and U.S. neglect. One factor stands out, however: the division of Latin America into two opposing factions. The first aligns itself generally with the "Washington Consensus," emphasizing representative democracy, the market economy, globalization, and cordial relations with the United States. The countries in this first group include Mexico, the Dominican Republic, Costa Rica, Panama, Colombia, Peru, Chile, Uruguay, and Brazil. The second faction favors participative democracy and rejects U.S. and capitalist models. It is led by Cuba and Venezuela, with important support in Mexico (López Obrador and the PRD), El Salvador (the FMLN), Nicaragua, Colombia (the FARC and part of the Polo Democrático), Ecuador, Bolivia, Argentina, and very probably Paraguay.

The divisions are not always neat: elements of the first faction can be found in the heart of the second, and, in turn, many of its governments are under siege by forces financed, organized, and trained by the other side. Some countries—Argentina and, to a lesser degree, Guatemala—shift between one side and the other. There is nevertheless a fundamental asymmetry between the two factions: the followers and practitioners of macroeconomic orthodoxy, of the democracy formerly called "bourgeois," and of understanding with Washington, even as governed by Bush, are timid, introverted, and cautious to an extreme (it was King Juan Carlos of Spain who famously told Chávez to "Shut up," not Felipe Calderón, Álvaro Uribe, Alán García, Michelle Bachelet, Tabaré Vázquez, or Lula). Neither do these leaders feel a pressing need to export or expand their "model": Brazil, for example, seeks greater influence in the region and in the world, but more for geopolitical motives than with any ideological bias. The other side, in contrast, has for many years had a strategy of expansion, and more recently has acquired the means to pursue it. Instead of Che Guevara's old dream of "one, two, many Vietnams," the rallying cry today might well be "one, two, many Venezuelas." The model here is winning power by the ballot, and conserving, transforming, and concentrating it via constitutional reform, armed militias, and monolithic parties, financed by revenue from the state oil company (PDVSA), defended and promoted by Cuban security squads, softened and sustained by social policies wrong over the long term but seductive in the short, and implemented by Cuban physicians, teachers, and instructors—all of it backed, in theory and increasingly in practice, with arms from Russia.

The second faction also has the advantage of a convincing discourse or narrative. In the context of persistent poverty and inequality, recurrent U.S. aggression and/or negligence, a venal private sector, and a history of corrupt and incompetent governments (AD and COPEI in Venezuela, the Colorados in Paraguay, Argentina's UCR, the Guayaquil oligarchy in Ecuador, the Santa Cruz economic class in Bolivia, etc.), the Bolivarian alternative seems like a dream come true. For a devastating diagnosis, it proposes an appealingly easy cure: education and health services delivered to the poorest sectors via so-called "missions" and Cuban cadres, paid for through nationalizations (Venezuela, Bolivia) or higher fees and taxes on the services of foreign companies (retentions on soy exports in Argentina, telecommunications and petroleum in Ecuador, electricity from the Itaipu and Yacyretá dams in Paraguay, gas pipelines and telecommunications in Bolivia). Under threat of expropriation, price reductions and controls are imposed on essential products such as cement, steel, flour, bread, etc. Best of all, everything is accomplished within a framework of democracy.

## Acting Abroad to Mask Trouble at Home

This discourse has become more aggressive in the past year in response to domestic pressures in both Cuba and Venezuela. For the Castro regime, the survival of the Chávez government is a matter of life and death. Without

Venezuelan subsidies, Cuba would be unable to pay for its basic oil needs (90,000 barrels per day), let alone be able to provide the population with the "modern" consumer products (cell phones, computers, DVD players) with which Raúl Castro seeks to buy time and patience to ease the Cuban transition. Cuba's leaders in turn are best able to help Chávez not on his most vulnerable flank—the domestic economy and political scene, with its outspoken opposition—but rather on the international front, contributing to his efforts to conquer abroad what he is losing at home. As Chávez has attempted to deal with the emergence of an opposition student movement, a disenchanted military sector (led by former defense minister Raúl Baduel), and defeat in the December 2007 constitutional referendum, Havana and Caracas have transformed their utopian Guevaran script into an immediate and pragmatic operating plan.

Here it is worth inserting a parenthesis to comment on Chávez's strategic defeat in that referendum, which generated many dubious or unsupported—though not necessarily false—assertions. What happened? The conventional theory that high abstention rates favor Chávez turned out to be wrong. At 44 percent, the abstention rate was close to the average in comparable elections over the last nine years. In a similar referendum in 1999, as many as 60 percent of voters stayed home. In 2007, however, the country did not polarize between the poor and the middle classes. Everything seems to indicate that an important part of Chávez's social base abandoned him.

Something happened between the time that the polls closed and the moment when Chávez acknowledged his defeat, but precisely what, we do not know. It seems unlikely that a Latin American president in full power and with a prior record of giving in to antidemocratic temptations would accept defeat by 1.4 percent of the votes without insisting on a recount. Losing candidates in both Costa Rica and Peru have done so recently, as did López Obrador in Mexico's presidential elections in 2006. A number of opposition leaders, as well as sources close to both Chávez and to his critics, have suggested explanations.

"That night, because of what we knew about certain meetings of which we were informed, I became very worried," the student leader Yon Goicoechea told the Spanish newspaper El País. "I cannot reveal the information. Some was confirmed and the rest not. . . . I would prefer not to comment on it." As to whether opposition leaders met with Chávez, he responded, "I imagine so; I don't know." General Baduel, in a CNN interview two days after the referendum, said that Chávez was determined to reject defeat within a margin of a few percentage points, but that the Venezuelan Armed Forces, "faithful to their democratic tradition," opposed an attempt to "act against the people," as in the failed April 2002 coup against Chávez himself.

From these reports and other speculation, the following scenario emerges. Before 6:00 P.M., based on his own exit polls and quick count, Chávez already knew that the "No" vote had won by four to six percentage points. He decided to refuse to accept this outcome, and to instead denounce a conspiracy, already named "Operation Pincers," supposedly engineered by "the forces of imperialism," the King of Spain, Álvaro Uribe, and all of Chávez's other bêtes noires. Four or five hours later, at around 11:00 P.M., the Armed Forces High Command urged him to accept the results or be removed. Baduel himself spoke by phone with Chávez at about 11:30 P.M. to tell him the same thing. Chávez allowed himself to be convinced, but asked in exchange that the opposition's official margin of victory be reduced to approximately one percentage point, allowing him to save face and present himself to the world as a magnanimous democrat. All that remained was to negotiate with members of the opposition to ensure that they would not publicize a different result once all of the electoral reports had been counted.

If this is true—and again we must recall that it is merely informed speculation—then Baduel would be reincarnating the role of Rear Admiral Wolfgang Larrazábal in 1958, at the fall of Marcos Pérez Jiménez, preparing the way for a version of "*chavismo* without Chávez." In effect, many members of the Venezuelan military High Command may have said to themselves, "What we like about *chavismo* is the possibility to enrich ourselves and become part of the *boliburguesía*, distribute some of the oil surplus to popular sectors, and be a bit more independent from Washington. Why should we fight with the Church, among ourselves, with King Juan Carlos, Uribe, Aznar, Fox, Toledo, and García, with Bush and the Brazilian Senate, with the students and CNN, and ally ourselves with Ahmadinejad and Castro to get what we want? Why not continue the first part and avoid the second? Preferably with Chávez, if he learns to quiet down, but without him if he doesn't."

For Chávez, the way to sidestep, resolve, and displace his growing internal difficulties was to focus instead on external victories—recurring, spectacular, and easily capitalized. Thus the cheap oil for Central America and the Caribbean, support for sympathetic candidates in other countries' elections (Paraguay, El Salvador) and, perhaps above all, the humanitarian agreement with Colombia which Chávez attempted to condition on international recognition of the FARC as a legitimate fighting force. The Bolivarian project, which began as a dream and an option, had become an urgent necessity in the face of internal commotion and adversity.

In this scenario, all would seem to point to Colombia as the jewel in the crown. The Cubans are too intelligent to believe that countries like Mexico or Brazil could fall under the FARC's "Strategic Plan," as Manuel Marulanda called it in the now infamous computers of Raúl Reyes. They are content to keep Calderón and Lula as their friends or accomplices and are not looking to replace them with unconditional allies (López Obrador or the left wing of Brazil's Workers' Party). They may covet Argentina and Peru, but the Cubans do not have powerful fifth columns there. The FARC, however, despite their supposed distancing from Havana and the reticence of the Polo Democrático to be Chávez's line of communication, continued to hope that, absent a viable successor to Uribe, they could transform the March 2010 elections into a significant advance. That is why so many of Chávez's actions are directed at Colombia: the negotiations to liberate the hostages; the support and advocacy for recognition of the FARC, not to mention financial support for the guerrillas; the geographic encircling of Colombia by Venezuela and Ecuador; and cooperation with the FARC to create and finance the Bolivarian Continental Coordinator throughout Latin America.

## How to React?

Since at least 2002 and the failed coup against him, and certainly since the failure of the civic and national oil company strike in January 2003, Chávez has made it clear that he will not respect the traditional rules of the game, either internally or externally, as he pursues his dream of twenty-first-century socialism in Venezuela and the rest of Latin America. He has said what he would do and has done what he said. He has used his gigantic river of petrodollars to fund opposition movements and governments in Mexico, Nicaragua, El Salvador, Colombia, Ecuador, Peru, Chile, Argentina, Paraguay, Bolivia, and even Brazil (the Movimento dos Sem Terra, or Landless Workers' Movement). Some of these financial contributions or subsidies have been productive, others not; but the intention is consistent and clear: if Washington has intervened in the region since at least 1836, why shouldn't Caracas do so as well?

At home, Chávez has warned that he will nationalize foreign companies in Venezuela that export too much, reduce supply, show excessive profits, or control strategic sectors. He initially agreed on indemnification, but soon began to proceed with greater rigor. We have already seen the nationalization of CANTV, Electricidad de Caracas, SIDOR, oil company ventures in the Orinoco Belt, and Cemex, and future targets include Bimbo, FEMSA, Polar, and Maseca. Most telling, after accepting defeat in the December 2007 referendum, Chávez vowed to proceed with the reforms voters had rejected, by whatever means necessary.

The dilemma for Latin American governments—as for Washington and Brussels—lies precisely in this violent asymmetry. Chavez rejects accepted internal democratic rules and the forms of international coexistence, while his adversaries and interlocutors continue to respect them. At the Rio Group Summit in Santo Domingo, at various meetings of the OAS, and at future summits, Calderón, Uribe, Lula, and Bachelet—in the absence of the King of Spain—have and will continue to remain mute in the face of Chávez's grandstanding, interruptions, and irreverence. Even after the evidence of Venezuelan backing for the FARC contained in Raúl Reyes's computers, these democratic leaders are unlikely to try to put a stop to Chávez's unorthodox antics.

The democratic governments of the region are somewhat justified in not knowing how to respond. How to play tennis when the opponent is playing football? How to respect rules which the adversary disrespects and rejects? How to find the right balance without committing the excesses of the opponent or overplaying one's hand? In short, what can the democratic governments of Latin America do about Chávez without sinking to the level of his verbal diatribes and imitating his methods?

As Chávez's strategy unfolds, with setbacks and advances, we will see how successful he is at filling the vacuum left by the other side. Unfortunately, we will also see the region's conflicts proliferate and accentuate because of the very nature of his strategy: a perpetual *fuite en avant*, employed externally to deal with unsolvable internal problems. One possible response could be for another regional power—Colombia, Mexico, Peru—to use international forums to denounce the FARC as a terrorist organization and achieve general condemnation of all those who support it, following the model of the UN Security Counsel Resolution against the authors of the September 11 attacks. Together with other measures, this could produce a feeling of isolation in

Chávez, the military High Command that remains faithful to him, and the *boliburguesía*. Steps like this one could eventually lead to the only exit in sight from the Latin American dilemma: for the Venezuelans themselves to put an end to Chávez's adventures by the means they consider most appropriate.

# Contributors

Ralph S. Clem is Professor Emeritus of Geography and International Relations at Florida International University, Miami.

Anthony P. Maingot is Professor Emeritus of Sociology and Anthropology at Florida International University, Miami.

Julia Buxton is Senior Research Fellow, Centre for International Cooperation and Security, University of Bradford, United Kingdom.

Jorge G. Castañeda is Global Distinguished Professor of Politics and Latin American and Caribbean Studies at New York University and formerly Foreign Minister of Mexico.

Javier Corrales is Associate Professor and Chair of Political Science at Amherst College, Amherst, Massachusetts.

Norman Girvan is Professorial Research Fellow at the Graduate Institute of International Relations at the University of the West Indies in St. Augustine, Trinidad and Tobago, and formerly Secretary General of the Association of Caribbean States.

John Magdaleno G. is Professor of Political Science at Universidad Simón Bolívar, Caracas, Venezuela.

Román D. Ortiz is Professor of Economics at Colombia's Universidad de los Andes and Director of Information and Analysis at Grupo Triarius, a Bogotá-based security and political risk consultancy firm.

María Teresa Romero is Associate Professor of International Studies at Universidad Central de Venezuela in Caracas.

Harold A. Trinkunas is Associate Professor and Chair of National Security Affairs at the Naval Postgraduate School in Monterey, California.

# Index

www.ingramcontent.com/pod-product-compliance
Lightning Source LLC
Chambersburg PA
CBHW031434270326
41930CB00007B/698